PHYSICS AND ASTRONOMY

DIMENSIONS OF SCIENCE
Series Editor: Professor Jeff Thompson

PHYSICS AND ASTRONOMY

Donald McGillivray

B.Sc., Cert. Ed.

MACMILLAN

First published 1987

Published by
MACMILLAN EDUCATION LTD
Houndmills, Basingstoke, Hampshire RG21 2XS
and London
Companies and representatives
throughout the world

Typeset by TecSet Ltd,
Wallington, Surrey
Printed in Hong Kong

British Library Cataloguing in Publication Data
McGillivray, D.
   Physics and astronomy. — (Dimensions of
   science)
   1. Metrophysics
   I. Title   II. Series
   523.01   QB461

ISBN 0-333-42861-7

To Elizabeth and Nicholas

Contents

Series Editor's Preface

This book is one in a Series designed to illustrate and explore a range of ways in which scientific knowledge is generated, and techniques are developed and applied. The volumes in this Series will certainly satisfy the needs of students at 'A' level and in first-year higher-education courses, although there is no intention to bridge any apparent gap in the transfer from secondary to tertiary stages. Indeed, the notion that a scientific education is both continuous and continuing is implicit in the approach which the authors have taken.

Working from a base of 'common core' 'A'-level knowledge and principles, each book demonstrates how that knowledge and those principles can be extended in academic terms, and also how they are applied in a variety of contexts which give relevance to the study of the subject. The subject matter is developed both in depth (in intellectual terms) and in breadth (in relevance). A significant feature is the way in which each text makes explicit some aspect of the fundamental processes of science, or shows science, and scientists, 'in action'. In some cases this is made clear by highlighting the methods used by scientists in, for example, employing a systematic approach to the collection of information, or the setting up of an experiment. In other cases the treatment traces a series of related steps in the scientific process, such as investigation, hypothesising, evaluating and problem-solving. The fact that there are many dimensions to the creation of knowledge and to its application by scientists and technologists is the title and consistent theme of all the books in the Series.

The authors are all authorities in the fields in which they have written, and share a common interest in the enjoyment of their work in science. We feel sure that something of that satisfaction will be imparted to their readers in the continuing study of the subject.

Preface

This book has been written for Advanced Level Students of Physics with a special interest in the Physics of Astronomy. It will also be a valuable text for Undergraduate Students who are studying astrophysics options in the first year of a degree course.

The purpose of the book is to show the application of physics to astronomy, and to draw attention to the advances that astrophysics has made during recent years.

I have used material which is close to the frontiers of knowledge on the subject, but at the same time I have tried to ensure that this material is clearly explained or reasoned.

I hope that the reader will find the book enjoyable and stimulating.

I should like to thank my colleague F. Starkey, and Dr Mykura of Warwick University, for their assistance, and Mrs J. Pope who made many helpful suggestions regarding the original manuscript; also Mrs B. Ward and Miss B Ward who typed the manuscript.

Donald McGillivray

Acknowledgements

Thanks are due to the following who have kindly permitted the reproduction of copyright photographs or diagrams

Figures 3.4, 7.6a, 7.10 and 9.3a The Royal Astronomical Society
Figures 7.7a and 7.7b   Photographs from the Hale Observatories
Figure 11.1   The Rutherford–Appleton Laboratory
Figures 11.2 and 11.4   National Aeronautics and Space Administration

1 Electromagnetic Radiation and Spectroscopy

ELECTROMAGNETIC WAVES

An electromagnetic wave is a flow of electromagnetic energy, consisting of mutually perpendicular electric and magnetic fields, which are oscillating transversely. Electromagnetic waves are generated whenever charged particles are accelerated; they require no material medium for their propagation, and have a constant velocity of 3×10^8 m s^{-1} in a vacuum. A diagrammatic representation of an electromagnetic wave is shown in figure 1.1.

The wavelength λ associated with any particular radiation is the distance between the two successive crests and since the velocity c of electromagnetic radiation is constant in a vacuum, the wavelength is determined by the frequency f of oscillations of the charged particle and is given by the equation

$$c = f\lambda$$

THE ELECTROMAGNETIC SPECTRUM

There exists a continuum of electromagnetic radiation, known as the *electromagnetic spectrum*. It includes in order of increasing frequency: radiowave, infrared, visible light, ultra violet, X-ray, gamma-ray and cosmic gamma radiation. Figure 1.2 shows the full electromagnetic spectrum, subdivided into intervals of wavelength and indicating the approximate limits of the various classes of radiation.

Above the range of wavelengths in metres is shown the corresponding range of frequencies in hertz. It should be noted that if the corresponding frequency above is taken for any wavelength below, then the product of frequency and wavelength is always 3×10^8 m s^{-1}. The 'visible light' region of the spectrum is relatively narrow and is expanded at the bottom of the diagram. Wavelengths for the visible region are given in 'nanometres'

1

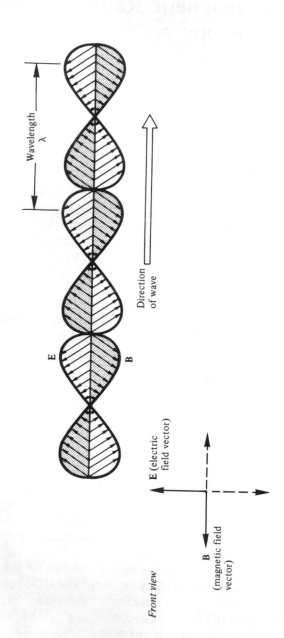

Figure 1.1 *Representation of an electromagnetic wave*

2

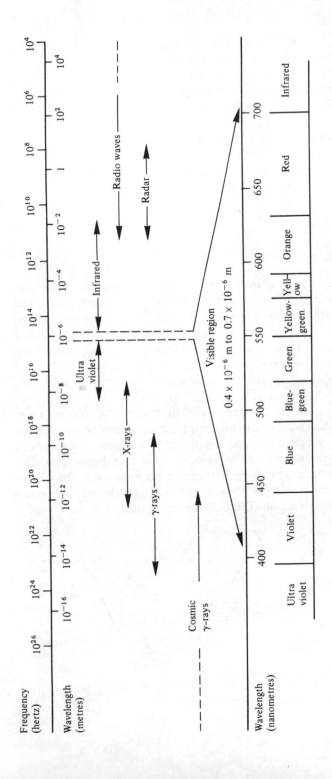

Figure 1.2 *The electromagnetic spectrum*

3

which are 'metres × 10^{-9}', since it is convenient to use the nanometre as a unit of wavelength in later sections. The wavelengths for visible light are also subdivided into colour regions, since the colour sensations which are produced in our minds are dependent upon the wavelengths of radiation which are received by our eyes; for example, all electromagnetic radiations with wavelengths in the range 630 to 700 nm appear to us to be 'red', and wavelengths from 520 to 550 nm appear 'green'.

The various categories of radiation are distinguished according to how the radiation is either produced or detected and therefore there are no distinct boundary lines, and for some categories there is considerable overlap.

DISPERSION

Stars are unique in being sources which emit the full range of wavelengths shown in figure 1.2. Other sources are restricted to a limited range of wavelengths, as for example a tungsten filament lamp, which gives a continuous band of radiation from the infrared, through the visible region and into the ultra violet. If we observe the light from a tungsten filament lamp with our eyes, the combined effect of the constituent wavelengths is to produce in our minds a sensation that we are perceiving 'white' light. This is a purely psychological effect.

A prism can be used to disperse 'white' light into its constituent wavelengths. Although all wavelengths of electromagnetic radiation travel at the same speed in a vacuum, they do not do so in a material medium, and shorter wavelengths are reduced in velocity more than longer ones. The amount of deviation produced by the prism is proportional to the relative decrease in velocity, and those wavelengths which are slowed down more undergo the greater deviation. This is illustrated in figure 1.3.

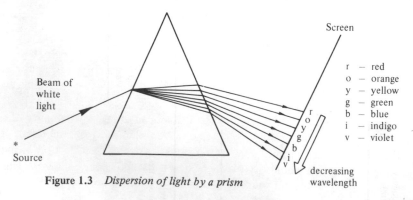

Figure 1.3 *Dispersion of light by a prism*

4

PRISM SPECTROMETERS

A device for analysing the constituent wavelengths of radiation emitted by a source is known as a *spectrometer*. A spectrometer which uses a prism to 'disperse' the various wavelengths is shown in figure 1.4.

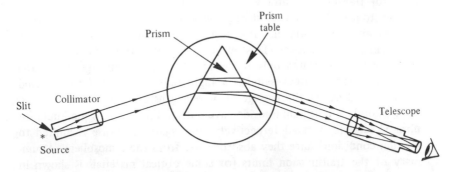

Figure 1.4 *A prism spectrometer*

Spectrometers of this basic design are suitable for examining wavelengths in the infrared, visible and ultra violet regions.

A narrow slit is illuminated by the light source and a converging lens at the end of the collimator is used to make the beam of light 'parallel'. The parallel beam is incident on one face of the triangular prism which deviates each wavelength by a different amount, so that an infinite number of parallel emergent beams are formed. These beams fall on to the objective lens of a telescope and are converged so that any point in the focal plane is the focus for one wavelength of radiation only. A spectrometer therefore produces a series of images of the illuminated slit and the image for each wavelength present is located at a different position in the field of view of the telescope. If a tungsten filament lamp is used as the source, then a 'pure' spectrum is formed with no overlapping of wavelengths. A photographic plate may be mounted in the focal plane of the telescope to produce a permanent record in the form of a 'spectrogram'.

When starlight is being analysed by a spectrometer the collimator is superfluous since the beam of light is already parallel, having travelled so far through space. The slit too is sometimes dispensed with since the star is a point source of light. However, use of a slit can improve the quality of a spectrogram in some cases. In order to 'widen' the spectrum, the image of the star is usually permitted to 'drift' at right angles to the direction of the dispersion. With a laboratory source of light, absence of the collimator lens or having too wide a slit produces a blurred final spectrum.

The use of prisms in spectroscopy is limited, because of the absorption of radiation by the material from which the prism is made. Glass prisms, for example, absorb strongly in the ultra violet and infrared regions. Quartz, rock salt and fluorite prisms and lenses are preferred for work in these regions, being much less absorptive than glass. Fluorite is especially good for transmitting ultra violet, and will allow wavelengths down to 125 nm to pass. However, air strongly absorbs ultra violet radiations below 180 nm and laboratory investigations below this wavelength must be undertaken in a vacuum. Low wavelength ultra violet radiation from stars can be investigated only if the spectrometer is operated above the earth's atmosphere. Rock salt has a high transmission limit for infrared radiation and allows the passage of wavelengths up to 14 500 nm. Potassium chloride and lithium fluoride have outstanding transmission limits for ultra violet and infrared respectively, but again their use is limited to vacuum conditions since they absorb water from the atmosphere. A summary of the transmission limits for some optical materials is shown in figure 1.5.

Material	Transmission limits (nm)	
	ultra violet	infrared
Crown glass	350	2, 000
Flint glass	380	2, 500
Quartz	180	4, 000
Fluorite	125	9, 500
Rock salt	175	14, 500
Potassium chloride	180	23, 000
Lithium fluoride	110	7, 000

Figure 1.5   Transmission limits for some optical materials

DIFFRACTION GRATING SPECTROSCOPY

A spectrum can also be formed using a 'diffraction grating'. There are two types, known as 'transmission' gratings and 'reflection' gratings, and either can be mounted on a spectrometer in place of the prism. When a transmission grating is used, the light under analysis is shone into the back of the grating and is dispersed as it passes through; with a reflection grating, light is shone on to its surface and dispersed as it is reflected. The two types of arrangement are shown in figure 1.6. Either type of grating consists of a sheet of glass or celluloid which is ruled with many fine, parallel and

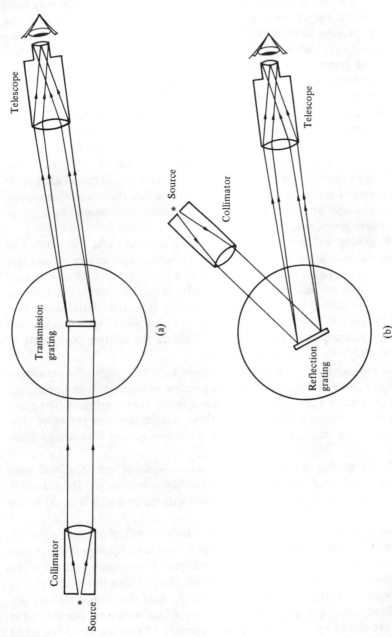

Figure 1.6 *Showing the arrangements for (a) a transmission grating, and (b) a reflection grating*

closely spaced grooves. For a reflection grating the grooves may be ruled on a polished metal surface using a point. In some gratings there may be as many as 20 000 of these per centimetre, and the spacing between the grooves may be as close as about 2 or 3 wavelengths of the light passing through. The actual ruled surface may be up to 20 cm long and 15 cm wide.

In the transmission grating the grooves scatter the incident light and are essentially opaque, while the unmarked parts of the surface transmit light. With the reflection grating the grooves scatter the light while the unmarked parts reflect the light regularly. In either case light emerges as if from a series of long, parallel, finely spaced slits. As the light emerges from these slits it fans out in all directions. The waves of light from different 'slits' interfere with each other in such a way that at any given point beyond the grating, the waves interfere destructively and cancel each other out, except for one particular wavelength for which there is reinforcement. The telescope brings into focus the successive wavelengths of light at successive positions in the focal plane.

A grating will produce more than one spectrum and the number of these depends upon the slit spacing. If a 'white' light source is used, the centre of the field of view is occupied by a white line and the spectra may be observed by rotating the spectrometer telescope around the table on which the grating is mounted. Moving away from the central white line, the successive spectra are known as the first order spectrum, second order spectrum and so on. Figure 1.7 shows the relative positioning of these spectra for a transmission grating.

A concave type of reflection grating is especially useful for ultra violet and infrared spectroscopy, since it is possible to mount this grating in such a way that the collimator and telescope lenses are superfluous. The transmission problems encountered at these wavelengths are therefore sidestepped. The 'Paschen' mounting for a concave grating is shown in figure 1.8.

If the grating is ruled on a concave cylindrical mirror, it will both disperse and focus the light. In the Paschen mounting the spectra are in focus around a circle and this circle has a diameter which is equal to the radius of curvature of the concave grating.

In addition to the advantages that a concave reflection grating has for ultra violet and infrared work, gratings have other advantages over prisms. 'Resolving power' is the ability to distinguish very close wavelengths and is usually greater for a grating than a prism. The resolving power of a grating increases with the total number of rulings. Also the wavelengths are distributed evenly across a grating spectrum, whereas they are compressed at the red end of a prism spectrum. The intensity of a grating spectrum might be less than that of a prism spectrum, since with a grating a given amount

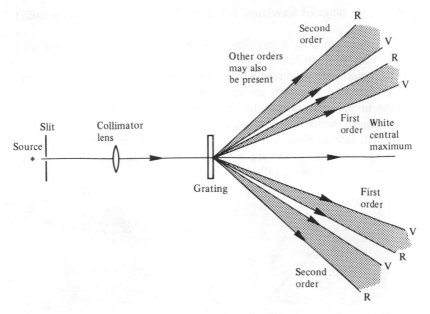

Figure 1.7 *Diagrammatic representation of orders of spectra from a grating*

of light energy is shared between a number of spectra. However, this problem can be overcome by shaping the profile of each groove so that the light is concentrated into one particular order.

TYPES OF SPECTRA

There are two categories of spectra known as 'emission' spectra and 'absorption' spectra. Emission spectra are observed when a source is examined directly by a spectrometer. Absorption spectra are observed after radiation from a source has passed through a substance, usually a gas, before it is examined by a spectrometer. Examination of an emission spectrum is concerned with looking at which wavelengths are present, whereas examination of an absorption spectrum is concerned with finding which wavelengths are absent after having been absorbed by the substance. With emission spectra the wavelengths that are present characterise the source, whereas with absorption spectra the wavelengths which are absent characterise the substance that the radiation has passed through. When an absorption spectrum is being produced, light from a 'continuous' source, giving a full range of wavelengths, is passed through the substance under investigation.

Both emission and absorption spectra can be further classified into either 'line', 'band' or 'continuous' spectra.

Line spectra are produced in low density monatomic gases and vapours at high temperatures, with atoms so far apart that they do not interact. Sources of line spectra include low pressure discharge tubes such as the sodium lamp, flames, and gaseous nebulae which are remains of exploded stars. Such sources emit only certain discrete wavelengths of radiation and produce a spectrum consisting of a number of separated images of the illuminated slit. These are called *spectral lines*.

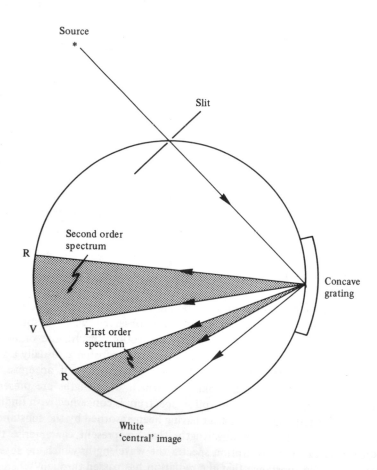

Figure 1.8 *A concave reflection grating in a 'Paschen' mounting*

An emission line spectrum consists of several bright lines of definite wavelength separated by dark gaps. An absorption line spectrum consists of the reverse with several dark lines representing definite wavelengths superimposed on an otherwise continuous spectrum. A sodium vapour spectrum is characterised by two discrete wavelengths at 589 nm and 589.6 nm. An emission spectrum produced by a sodium lamp therefore consists of two bright yellow lines which are very close together, against a black background. An absorption spectrum of sodium vapour consists of a continuous band of colours which is interrupted by two black lines in the yellow region. The lines in each case correspond to the characteristic 'sodium vapour' wavelengths. These two types of line spectra are represented in figure 1.9.

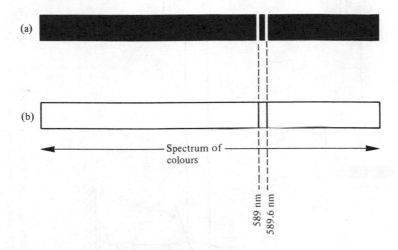

Figure 1.9 *(a) Line 'emission' sodium vapour spectrum. (b) Line 'absorption' sodium vapour spectrum*

Sodium vapour spectra are convenient to observe in a laboratory. The emission spectrum of sodium vapour can be observed if a sodium lamp is viewed directly through a spectrometer. A 'sodium lamp' operates by having an electrical discharge pass through sodium vapour at low pressure. The light produced is emitted by the atoms of sodium and has a characteristic yellow colour. One means of observing a sodium vapour absorption spectrum is to observe light from a tungsten filament lamp after it has passed through a sodium flame in a bunsen burner. These arrangements are shown in figure 1.10.

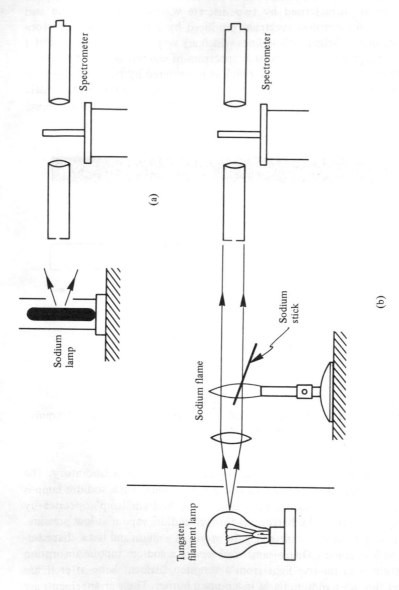

Figure 1.10 *(a) Observing a sodium vapour emission spectrum. (b) Observing a sodium vapour absorption spectrum*

Band spectra consist of a number of groups of closely spaced lines. Each group itself is well-defined and is known as a 'band'. Within each band the lines become more closely packed at the side representing shortest wavelength, so that this side appears to have a sharp and bright edge. Band spectra are produced by gases and vapours with polyatomic molecules such as oxygen, carbon dioxide and water vapour. Figure 1.11 illustrates how the spectral lines are grouped into bands.

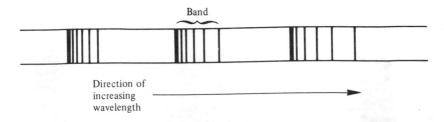

Figure 1.11 *Showing the arrangement of lines in a band spectrum*

The importance of line and band spectra is due to the fact that each element gives rise to a unique set of lines, and spectroscopy is a powerful tool for chemical analysis. The line spectrum of any substance is the sum of the line spectra of its constituent elements, and spectral analysis of chemical substances has proved far more accurate than any chemical methods of analysis. In astronomical spectroscopy the absorption lines present in the continuous spectra of stars enables the chemical composition of the stars to be identified.

Band spectra are more complex than line spectra because of interactions which take place between adjacent atoms within each molecule. Continuous spectra are produced in a solid, liquid, or very dense gas where the possibilities for such interactions are so increased that all wavelengths are generated. These interactions are a 'quantum' phenomenon and the origins of the different types of spectra are explained more fully in chapter 2.

The relative intensities of the constituent wavelengths depend mainly upon the temperature of the emitting body but also upon the nature of its surface. At temperatures below about 1000°C most of the emitted radiation is in the infrared region of the electromagnetic spectrum, but at higher temperatures visible and ultra violet radiation are also produced. This accounts for the continuous spectrum which is obtained from the heated filament of a tungsten filament lamp which has been mentioned earlier. Electromagnetic radiation which is emitted by a body due to its temperature is known as *Thermal Radiation*. Thermal radiation is given more consideration in the next chapter and is mentioned again when we consider the production of radio waves by stellar sources in chapter 10.

2 Black Body Radiation and Quantum Physics

THE ABSORPTION AND EMISSION OF RADIATION

The ability of a body to radiate is closely related to its ability to absorb radiation, which is to be expected since any body which maintains a constant temperature is in thermal equilibrium with its surroundings, and therefore is absorbing energy at the same rate at which it is emitting energy. Good absorbers are therefore also good emitters. Matt black surfaces are the best absorbers and the best emitters of radiation, whereas polished silver surfaces are poor emitters and poor absorbers.

It is useful to consider a 'black body', which is the idealised concept of a body which absorbs completely all radiation which falls upon it. No perfect black body actually exists, although soot absorbs 95 per cent of incident radiation of infrared and visible wavelengths. The ratio of the energy absorbed to the energy falling on a surface is known as the *absorptivity* of the surface. Absorptivity varies between 0 for a perfect reflector and 1 for a black body.

Since a black body is a perfect absorber it must also be a perfect emitter, and the radiation which it emits is known as black body radiation.

The *emissivity* of a surface is defined as the ratio of the energy emitted by a surface to that emitted by a black body of the same area in the same time. Emissivity varies between 0 for a perfect reflector and 1 for a black body. The absorptivity of a body is always numerically equal to its emissivity.

The usefulness of introducing the idealised black body in a discussion of thermal radiation is that it becomes possible to ignore any of the surface properties of the radiator which might affect the manner in which it radiates. All black bodies behave identically.

An experimental black body can be manufactured from a hollow porcelain sphere which has a small hole in it. The inside walls are blackened with soot so that most of any radiation entering the hole is absorbed by the wall. The small amount of radiation which is not absorbed in the first encounter with the wall is mostly absorbed on being reflected to another part of the wall. After a few reflections virtually no radiation remains unabsorbed and there is little chance of any radiation escaping from the hole, which therefore acts as a black body. This is illustrated in figure 2.1.

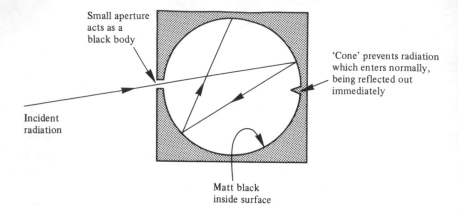

Figure 2.1   *An 'experimental' black body*

A perfect absorber is also a perfect emitter and a black body radiator can be made by surrounding the enclosure of figure 2.1 with a heating coil. It is then possible to examine the characteristics of the black body radiation which emerges from the hole with respect to the temperature of the enclosure. The radiation which is emitted by any section of the wall is involved in many reflections before it eventually emerges from the hole. Any part of the surface which is a poor emitter absorbs little of the radiation which falls upon it, while other parts which are good emitters absorb most of the radiation which falls on them. The net effect of this process is the complete mixing of radiations before they emerge, to make the temperature of the surface uniform.

The 'mixing' of radiations ensures that the radiation does not depend on any surface, and that the nature of the radiation emerging from the hole depends solely on the temperature of the surface.

BLACK BODY SPECTRA

An instrument for investigating the distribution of energy in the spectrum of black body radiation is known as a 'spectroradiometer' and is illustrated in figure 2.2.

In the spectroradiometer, radiation from a black body is collimated by a concave mirror and dispersed by a prism. The separated beams of radia-

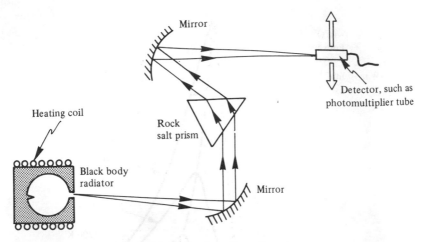

Figure 2.2 *A spectroradiometer*

tion are focused by a second concave mirror on to a detector. The detector can be moved along the spectrum to sample the energy which is being radiated at any wavelength. If the detector is a photomultiplier tube, then the output current from the photomultiplier is directly proportional to the radiant energy falling upon it. A photomultiplier tube is a suitable detector in this instance since its output does not vary with the wavelength of incident radiation, but only with the intensity of the radiation. Mirrors are used in preference to glass lenses which would absorb infrared radiation, and for the same reason the prism is made from rock salt. The spectroradiometer is used in astronomy to examine the energy characteristics of stellar spectra, in which case a telescope takes the place of the black body radiator and it is starlight which is dispersed by the prism.

Experimental results showing the way in which the energy of black body radiation is distributed along the spectrum of wavelengths are shown in figure 2.3.

The intensity of radiation is measured in terms of the *emissive power* (E_λ). The units of emissive power are Watts metre^{-2} nm^{-1}. The 'Watts metre^{-2}' term in these units refers to the energy radiated per second from a surface area of one square metre. The last term in the units, 'nm^{-1}', is included because it would be a nonsense to imagine that a single wavelength could be singled out from which to take measurements. Instead measurements are taken from a narrow band of wavelengths, and the width of the band is controlled by the width of the spectrometer slit. In this instance the band width (for measurement purposes) is chosen to be one nanometre.

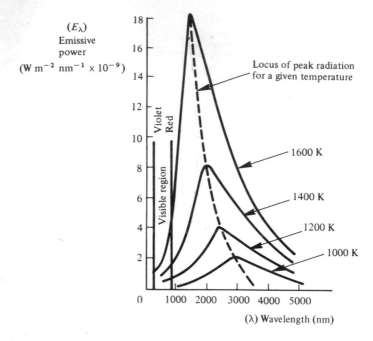

Figure 2.3 *Black body spectra*

Each curve in figure 2.3 corresponds to a certain temperature of the black body. It can be seen that as the temperature is increased, the intensity at each wavelength increases, but shorter wavelengths show the most marked increase. There is a position of maximum energy for each wavelength, located along the dashed line, and this maximum becomes displaced towards the shorter wavelengths as the temperature of the source increases.

None of the curves in figure 2.3 is emitting much, if any, radiation in the visible region, and if we extrapolate from the dashed line we see that we need very high temperatures for this to occur. The temperature of the filament of a tungsten lamp for example, which radiates all visible wavelengths, is about 3000 K. The curves also illustrate the well-known phenomenon that the colour of a body, which is hot enough to be radiating visible light, depends upon its temperature. At around 1200 K the visible radiations are confined to red wavelengths, but shorter visible wavelengths are emitted as the temperature increases so that the colour will change to yellow and eventually white when all visible wavelengths are emitted. It is common to refer to 'red heat' or 'white heat'. Even more intense heating produces a blue colour as the intensity of blue wavelengths increases until this colour predominates.

Wien's Law

Wien's law describes how the displacement of the peak of maximum radiation illustrated in figure 2.3 varies with the temperature of the black body. The relationship is

$$\lambda_{max} T = \text{constant}$$

The wavelength at which the peak of radiation occurs, multiplied by the temperature in Kelvin, is a constant. The value of Wien's constant is 2.9×10^{-3} m K.

The Use of Wien's Law to Find the Temperature of a Star

Stars themselves approximate to black body radiators, and the wavelength which corresponds to the peak power output provides an approximate method for determining the effective surface temperature T_{eff} of a star. For example the peak wavelength of the sun is 490 nm, from which the effective surface temperature is calculated as

$$T_{eff} = \frac{2.9 \times 10^{-3}}{490 \times 10^{-9}} = 5920 \text{ K}$$

that is, the effective surface temperature of the sun is 5920 K. It should be noted that there is a considerable variation in temperature throughout the various layers of a star (see page 38). The corona of the sun, for example, which does not radiate at visible wavelengths is at a temperature of 10^6 K, while the temperature of the thermonuclear core of the sun is 1.4×10^7 K. The effective surface temperature of a star is the temperature the stellar surface would have if it were a black body with the same surface area and luminosity.

The 'Colour' of a Star

Even to the naked eye the 'colours' of some bright stars can be perceived. Sirius, the brightest star in the sky, is very hot and appears blue. Betelgeuse in the constellation Orion is known as a 'red giant', and being relatively cooler appears red. Our sun appears yellow, since the peak wavelengths in the blue-green region combine with the longer wavelengths from the yellow, orange and red parts of the spectrum, to produce the overall yellow effect.

Stefan's Law

Stefan's law states that the total energy radiated by a black body in unit time per unit surface area is proportional to the fourth power of the temperature of the body expressed in Kelvin, that is

$$E = \sigma T^4$$

where E is the energy radiated per second per square metre of surface, T is the temperature of the black body in Kelvin, and σ is 'Stefan's constant' which has the value of 5.67×10^{-8} W m^{-2} K^{-4}.

A value for E at any temperature T, may be found from the curves of figure 2.3 since E is equal to the area under the corresponding curve. The area under each curve is given by

$$E = \int_0^\infty E_\lambda \, d\lambda$$

where E has the units W m^{-2}.

Since a body which is radiating energy is also receiving energy, the net loss of energy depends upon the temperature difference between the body and its surroundings. If the temperature of the body is T and the temperature of the surrounds is T_0 then the energy radiated from the body per second, per square metre, is given by σT^4, and the energy received by the body per second, per square metre is σT_0^4. Hence the net loss of energy per second per square metre is given by

$$E_{\text{net}} = \sigma \left(T^4 - T_0^4 \right)$$

If the body concerned is not a true black body then Stefan's law may be modified using the 'emissivity' constant which was introduced earlier, and has a value between 0 for a perfect reflector and 1 for a perfect black body. Stefan's law becomes

$$E = \epsilon \sigma T^4$$

and the net loss of energy per second per square metre to the surroundings becomes

$$E_{\text{net}} = \epsilon \sigma \left(T^4 - T_0^4 \right)$$

The Use of Stefan's Law to Calculate the Effective Surface Temperature of the Sun

In the following section we make the approximation that both the earth and the sun are black body radiators.

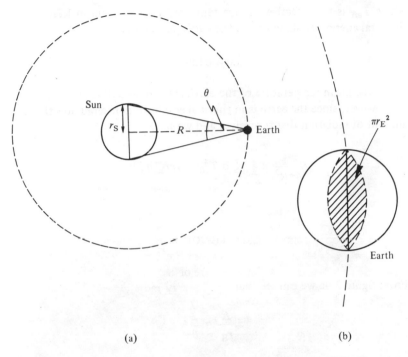

(a)   (b)

Figure 2.4   *The use of Stephan's law to calculate the effective surface temperature of the sun*

Consider figure 2.4a; R is the radius of the earth's orbit, r_E is the radius of the earth and r_S is the radius of the sun.

The fraction of the sun's energy which is received by the earth is given by the ratio of the earth's diametral plane, which is the shaded area shown in figure 2.4b, to the area of the sphere of radius R surrounding the sun. Hence the fraction of the sun's energy received by the earth is

$$\frac{\pi r_E^2}{4\pi R^2}$$

From Stefan's law we have that the energy radiated per second per square metre from the sun is

$$\sigma\, T_{\text{eff}}^4$$

and the total energy radiated from the sun per second is therefore

$$4\pi r_S^2\ \sigma\, T_{\text{eff}}^4$$

where T_{eff} is the effective surface temperature of the sun in Kelvin. Also the total energy reradiated from the earth per second is

$$4\pi r_E^2\ \sigma\, (290)^4$$

where the mean temperature of the earth is taken as 290 K (17°C).

However, since the earth is in thermal equilibrium it reradiates the same amount of radiation that it receives and we have

$$\frac{\pi r_E^2}{4\pi R^2} \times 4\pi r_S^2\ \sigma\, T_{\text{eff}}^4 = 4\pi r_E\ \sigma\, (290)^4$$

whence

$$T_{\text{eff}}^4 = 4\,(290)^4 \times \frac{R^2}{r_S^2}$$

From figure 2.4a we can see that

$$\frac{R}{r_S} = \frac{2}{\theta}$$

Where θ is the angle subtended at the earth by the diameter of the sun, and is approximately 0.0094 radians. Therefore

$$T_{\text{eff}} = 4\,(290)^4 \times \frac{4}{\theta^2} = \frac{(290)^4 \times 16}{(0.0094)^2} = 6010 \text{ K}$$

In this way the temperature of the sun is found, but we note that the value found using Stefan's formula is at variance with the result we had using Wien's law. However, both methods can be judged as only approximations, since the assumption that the sun is a black body is made in both examples and this is not completely justified. In the last example we even assumed that the earth was a black body, which makes the last answer very approximate. The earth also receives heat from its interior which is another factor that has been ignored.

The Total Radiative Power Output of a Black Body

If E is the energy radiated per unit area per second, then the total radiative power output of a black body is given by

$$P = AE = A \, \sigma \, T^4$$

where A is the total surface area of the black body and P is measured in Watts.

The 'Luminosity' (L) of a Star

The luminosity of a star is defined as the total power radiated by the star at all wavelengths. Luminosity is measured in Watts.

The power output from each square metre of the surface of a star is given by Stefan's law

$$E = \sigma \, T_{\text{eff}}^4 \; (\text{Watts m}^{-2})$$

The luminosity of a star is therefore given by the product of the surface area of the star and the quantity '$\sigma \, T_{\text{eff}}^4$', that is

$$L = 4\pi R^2 \; \sigma \, T_{\text{eff}}^4$$

QUANTUM PHYSICS

Towards the end of the nineteenth century a number of attempts were made to deduce mathematical equations which would represent the black body radiation curves shown in figure 2.3. The most notable of the attempts that 'failed' was made by Rayleigh and Jeans who derived an equation based upon an assumption that the radiation inside a black body cavity was reflected within the cavity in the form of a three-dimensional, standing electromagnetic wave. Although their reasoning was sound, their final formula did not fit the observed facts.

The Rayleigh-Jeans formula gave a good 'fit' with the black body radiation curves at long wavelengths but was in complete disagreement in the ultra violet region. The failure of the Rayleigh-Jeans formula became known as the 'ultra violet catastrophe'. It marked the end of the period known as 'classical' physics and marked the beginning of 'quantum' or 'modern' physics.

In 1900 Max Planck derived, by trial and error, a formula which fitted the radiation curves exactly. Planck's radiation formula states that

$$E_\lambda = \frac{8\pi ch}{\lambda^5} \times \frac{1}{(\exp(ch/k\lambda T) - 1)}$$

where E_λ is the emissive power for a given wavelength λ, T is the temperature in Kelvin, k is Boltzmann's constant, c is the velocity of electromagnetic radiation and h is 'Planck's constant' and has a value of 6.62×10^{-34} Joule seconds.

Planck then attempted to derive his formula from first principles and began with the assumption that the inside wall of a black body cavity contained 'electrical oscillators' (vibrating electrons perhaps) which radiated electromagnetic waves. Initially, Planck had no reason to doubt that his oscillators could emit energy continuously, but he found he could only derive his formula if he made the assumption that his oscillators emitted radiation discontinuously, as discrete bursts of energy. He called these bursts of energy *quanta*. Planck showed that the quanta associated with a particular frequency f of radiation, all have the same energy E, and also that this energy is directly proportional to f so that

$$E = hf$$

where h is Planck's constant. It follows that a 'quantum' of high frequency represents a larger amount of energy than a quantum of low frequency.

Planck's assumption that the energies of his oscillators were 'quantized' was completely radical and was treated with scepticism. Planck himself tried to find alternative explanations. However, in 1905 Albert Einstein took up Planck's quantum theory and developed it to explain the newly discovered 'photoelectric' effect, and in 1913 Niels Bohr used the quantum theory to account for the origin of spectral lines emitted from a hydrogen atom. Planck's assumption is now recognised as being among the most important discoveries in physics. A quantum of electromagnetic radiation is known as a *photon*.

The Electron Volt

The S.I. unit of energy is the Joule but since the energies associated with quanta are very small, the electron volt, eV, is a much more convenient unit of energy. The electron volt is defined as the amount of energy an electron acquires if it is accelerated between two points which have a potential difference of one volt.

Figure 2.5 shows an electron being accelerated between two plates which are at a potential difference of 1 volt. From the definition of the electron volt we see that the electron has acquired 1 eV of energy when it

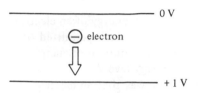

Figure 2.5 *Illustrating the 'electron volt'*

reaches the positive plate. To calculate the energy acquired by the electron in Joules we use the formula

$$\text{energy} = Q \times V$$
$$(\text{Joules}) = (\text{coulombs}) \times (\text{volts})$$

The charge on the electron is 1.6×10^{-19} coulombs, and so the energy acquired by the electron is given by

$$\text{energy} = 1.6 \times 10^{-19} \times 1 = 1.6 \times 10^{-19} \text{ Joules}$$

So we see that 1 eV is equivalent to 1.6×10^{-19} Joules which gives us the conversion factor. Care must be taken to be consistent if the electron volt is used as the unit of energy.

The Origins of Spectra

In 1913 Niels Bohr used 'quantum' ideas to construct a model of the hydrogen atom. This model had notable success in explaining the origins of the line spectra produced by hydrogen. Bohr's model also provided the basis for further models which led to an understanding of the production of line spectra of all the elements and not just hydrogen.

Bohr made two postulates concerning the orbiting electrons of atoms, both of which he could justify only on the grounds that the results he obtained by making them fitted exactly with the experimental evidence available.

Bohr's *first postulate* was that the electrons in any atom can exist only in certain discrete orbits, and while they remain in these orbits the electrons do not radiate energy.

In such an orbit an electron would possess some kinetic energy due to its motion and some potential energy due to its proximity to the positively

25

charged nucleus, and the total energy of an electron is therefore characterised by the radius of its orbit. Each permitted orbit would therefore have
associated with it a particular amount of energy and the various orbits can
be considered as being 'energy levels'.

Bohr's *second postulate* was that an electron may 'jump' from a higher
energy orbit to a lower energy orbit, and in doing so emit a single quantum
of radiation. The energy of this quantum is equal to the difference in
energies associated with the two orbits concerned. If E_2 is the energy of
the higher orbit and E_1 is the energy of the lower orbit, then the energy of
the emitted quantum is given by

$$
\underset{\substack{\left(\text{energy of} \atop \text{quantum}\right)}}{hf} = \underset{\substack{\left(\text{difference between} \atop \text{energy levels}\right)}}{E_2 - E_1}
$$

Bohr's postulates therefore describe an atom which is normally stable
and emits no radiation until some disturbance causes a 'quantum jump',
whereupon radiation of a discrete wavelength is emitted. Transitions of
this sort between different orbits give rise to the emission of a series of
discrete wavelengths which is detectable as a line spectrum.

The possible orbital energy states which are permitted to an electron
can be shown on an 'energy level' diagram. Bohr developed his theory
quantitatively for the hydrogen atom and obtained the energy level
diagram shown in figure 2.6. The diagram indicates the energies associated
with each permitted orbit as calculated by Bohr.

Each energy level is denoted by a quantum number n. The level at
which an electron has lowest energy is given by $n = 1$ and is known as the
'ground state'. If an electron is provided with extra energy it may go into a
higher energy level, in which case we describe the atom as being in an
'excited' state. Given enough energy the electron can escape entirely from
the atom, which corresponds to a quantum state of infinity. An atom
which has lost an electron is said to be 'ionised'. By convention we say
that an electron removed to infinity has zero energy, and if we maintain
that an electron loses energy when it 'falls' into an atom, the energy it
possesses at any quantum level must be less than zero, and so the energy
levels are assigned negative values. The ground state for hydrogen has an
energy value of -13.6 eV, and so to ionise the atom 13.6 eV of energy
must be supplied from outside. The amount of energy required to ionise
an atom is known as the *ionisation energy*.

At atom in an excited state is unstable and after a short, but indeter-
minate period, the electron falls back into a lower energy state with the
emission of radiation.

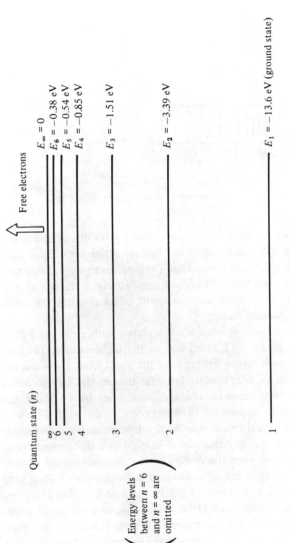

Figure 2.6 *Energy level diagram for the hydrogen atom*

Quantum state (n)

Free electrons

$E_\infty = 0$
$E_6 = -0.38$ eV
$E_5 = -0.54$ eV
$E_4 = -0.85$ eV

$E_3 = -1.51$ eV

$E_2 = -3.39$ eV

$E_1 = -13.6$ eV (ground state)

Energy levels
between $n = 6$
and $n = \infty$ are
omitted

The Spectrum of Atomic Hydrogen

Bohr was familiar with the spectrum of atomic hydrogen which consists of distinct groups of lines, as illustrated in figure 2.7.

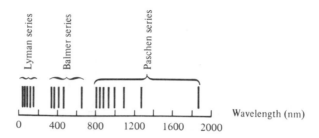

Figure 2.7 *Schematic diagram of the spectrum of the hydrogen atom*

Each group of lines is named after the person who discovered them. The first group to be discovered were the Balmer series since these lines occur at visible wavelengths. The Lyman and Paschen series were discovered later in the ultra violet and infrared regions respectively. Bohr was able to show that each one of these lines was the result of an electron transition between the various quantum states.

The Lyman series of spectral lines are associated with electrons falling back to the ground state (n = 1), from any of the higher states. The first Lyman line is due to an electron falling into the atom from n = ∞ when a photon of energy 13.6 eV is released. The last line in the Lyman series results from a transition, between the quantum level (n = 2) and the ground state, whereupon a photon of energy $E_2 - E_1$ = 10.21 eV is released. Similarly the penultimate line in the Lyman series corresponds to a transition between the quantum level (n = 3) and the ground state, where $E_3 - E_1$ = 12.09 eV, gives the energy of the photon emitted.

Lines of the Balmer series are all associated with electrons falling into the n = 2 quantum level, and those of the Paschen series, with electrons falling into the n = 3 quantum level. Other series of lines with longer wavelength have been identified and are associated with the other quantum states. The main electron transitions for atomic hydrogen are superimposed on the energy level diagram in figure 2.8.

Using the energy values of the quantum states which Bohr calculated from his theory, it is possible to calculate the wavelength of any spectral line using the formula

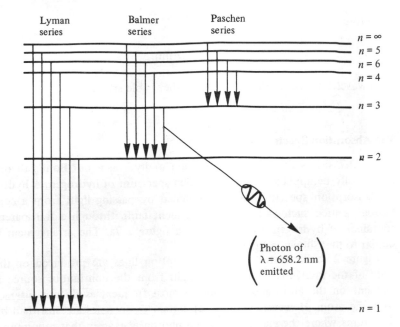

Figure 2.8 *Electron transitions giving rise to spectral lines for atomic hydrogen*

$$hf = \frac{hc}{\lambda} = E_n - E_n'$$

where E_n and E_n' are respectively the high and low energy levels involved.

The last Balmer line, for example, corresponds to a transition between the quantum levels $n = 3$ and $n = 2$, and the calculation is as follows

$$E_3 - E_2 = (-1.51) - (-3.39) = 1.88 \text{ eV}$$

Converting electron volts to Joules we have

$$E_3 - E_2 = 1.88 \times 1.6 \times 10^{-19} = 3.008 \times 10^{-19} \text{ J}$$

The wavelength of the spectral line is given by

$$\lambda = \frac{hc}{(E_3 - E_2) \text{ Joules}} = \frac{6.62 \times 10^{-34} \times 3 \times 10^8}{3.008 \times 10^{-19}}$$

therefore

$$\lambda = 658.2 \text{ nm}$$

This wavelength lies in the 'red' part of the visible spectrum.

The Absorption Spectrum of Hydrogen

Figure 2.7 shows the emission spectrum of hydrogen but Bohr's theory can equally be applied to the absorption spectrum of hydrogen. A hydrogen absorption spectrum can be obtained by passing light from a continuous source such as a tungsten filament lamp through a transparent container of hydrogen gas, as shown in figure 2.9a. The arrangement is similar to that shown in figure 1.10b.

Figure 2.9b illustrates how the absorption lines are produced on the scale of the individual atom. White light from the continuous source is incident on the atom and photons of most frequencies do not interact with the atom. However, photons of certain frequencies are absorbed by the atoms where the energy of such a photon is exactly that required to transfer an electron between energy levels in the atom. The excited electrons fall back to lower energy levels and reradiate photons of the same frequencies as those that were absorbed, but the reradiation now occurs in all directions, and so the quantity of these photons travelling in the original direction is greatly depleted. The extent of the depletion is such that these frequencies are so faint that they appear as dark lines on the continuous background, giving the absorption spectrum.

Stellar spectra are characterised by thousands of absorption lines. These are discussed in chapter 3. In the case of the sun they are known as *Fraunhofer lines*.

The Wave-mechanics Model of Atomic Structure

Bohr's model of atomic structure which could only account for atomic hydrogen has now been superseded by the wave-mechanics model, which can account for the spectra of all elements. This model is however mathematical and complex. The electrons are given wave properties and rather than being thought of as having definite orbits we consider the 'probabilities' of an electron being found at a given distance from the nucleus. The notion of there being discrete energy states, and the transitions between them giving rise to photons, is however retained.

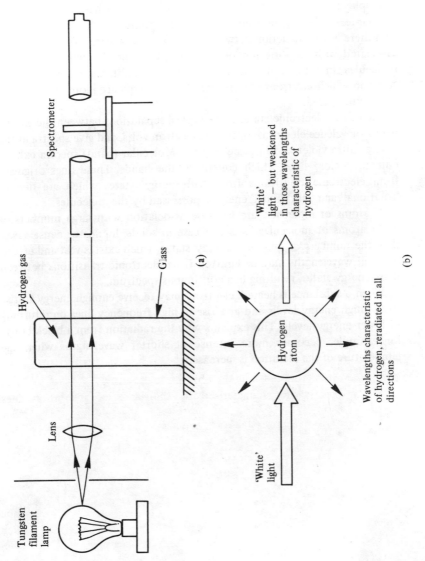

Figure 2.9 (a) Obtaining an absorption spectrum of hydrogen gas. (b) Illustrating the production of absorption lines

Labels in figure (a): Spectrometer; Hydrogen gas; Glass; (a); Lens; Tungsten filament lamp

Labels in figure (b): 'White' light — but weakened in those wavelengths characteristic of hydrogen; Hydrogen atom; Wavelengths characteristic of hydrogen, reradiated in all directions; 'White' light; (b)

The Production of Band and Continuous Spectra

So far we have shown only how line spectra are produced, but the same principles apply with band and continuous spectra. As has been mentioned in chapter 1, line spectra are produced by atoms which are separated so that there is no interaction between them. In cases where atoms are closely associated with one or a few other atoms, as is the case with polyatomic molecules, then the atoms interact to produce additional quantum energy states in which electrons may reside. As a consequence more spectral lines may be produced.

Molecular electronic states have typical separations between the energy bands of valence electrons, of several electron volts, and give spectra in the visible, ultra violet and infrared regions. Molecular (band) spectra exhibit numerous close lines which constitute the bands. These lines originate from electronic transitions from 'sub-energy' states which are due to 'rotational' and 'vibrational' energies possessed by the molecules.

If atoms or molecules are in close association with large numbers of other atoms or molecules as is the case in solids, liquids or dense gases, then the numbers of quantum energy states which exist is vast and effectively all wavelengths may be emitted from electronic transitions between these energy states, resulting in a continuous spectrum.

When a substance is heated, electrons may receive enough energy to rise to the high energy levels and give rise to high frequency lines on returning to lower energy levels. This explains why the radiation from a black body, for example, becomes more intense at shorter wavelengths when the temperature of the radiation is increased.

3 Atmospheric Absorption of Radiation

ABSORPTION BY THE EARTH'S ATMOSPHERE

Although the stars emit radiation of all wavelengths, most regions of the spectrum cannot be observed from the earth's surface because of absorption by our atmosphere. All parts of the spectrum are affected to some extent, but for some regions of wavelength the atmosphere is completely opaque, even when observations are made from high mountain tops. Two regions of the electromagnetic spectrum to which the atmosphere is particularly transparent are the 'visible' region and a 'radio' region with wavelengths between 1 cm and 100 m. These two transparent regions are often referred to as the *optical window* and the *radio window* respectively. The extent of atmospheric absorption across the electromagnetic spectrum is illustrated in figure 3.1.

Gamma, X-rays and ultra violet radiations are absorbed by molecules of oxygen and nitrogen at height above 100 kilometres. Ozone is particularly responsible for the absorption of ultra violet, while absorption in the visible, infrared and radar regions is due mainly to water vapour and carbon dioxide.

Long range radio waves are not in fact absorbed by the atmosphere but are reflected back into space by the 'Ionosphere', which consists of layers of heavily ionised air at heights between 100 and 400 kilometres.

Details of the effects of atmospheric absorption on the spectrum of the sun are shown in figure 3.2.

Three curves are represented in figure 3.2. Curve B shows the sun's energy spectrum as it appears when observations are made above the atmosphere. Curve C is the theoretical radiation curve for a black body at a temperature of 6000 K, which is the approximate temperature of the surface of the sun. Curve A shows the sun's energy spectrum as recorded at sea level, and comparison with curve B reveals the effects of atmospheric absorption. The pronounced dips in curve A are due to absorption at specific wavelengths by those molecules which predominate in the atmo-

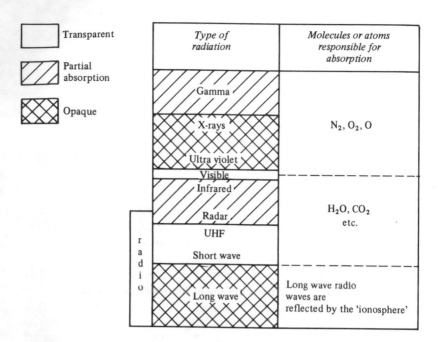

	Type of radiation	Molecules or atoms responsible for absorption
▢ Transparent	Gamma	
▨ Partial absorption	X-rays	N_2, O_2, O
▧ Opaque	Ultra violet	
	Visible	
	Infrared	
	Radar	H_2O, CO_2 etc.
	UHF	
	Short wave	
radio	Long wave	Long wave radio waves are reflected by the 'ionosphere'

Figure 3.1 *Showing the extent of atmospheric absorption across the electromagnetic spectrum*

sphere, such as water vapour, oxygen, ozone and carbon dioxide. These molecules produce 'band' spectra.

THE 'SCATTERING' OF LIGHT IN THE EARTH'S ATMOSPHERE

If the size of a particle in the atmosphere has the same order of magnitude as the wavelength of radiation passing through the atmosphere, then diffraction occurs around the edge of the particle and the radiation is 'scattered'. Scattering occurs around all atoms, molecules and dust particles in the atmosphere.

In 1871, Lord Rayleigh found that the extent to which light is scattered depends upon wavelength, such that the intensity of scattered light is inversely proportional to the fourth power of wavelength. That is

$$\text{Intensity of scattered light} \propto \frac{1}{\lambda^4}$$

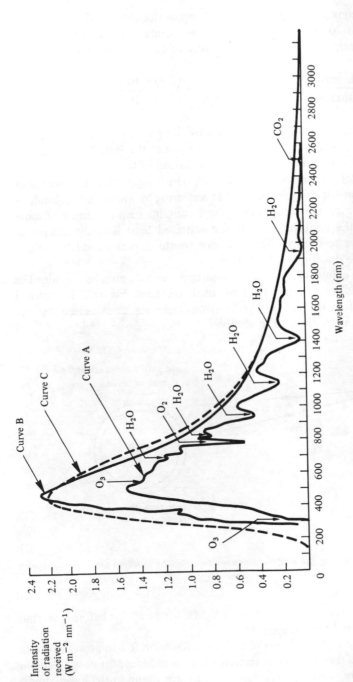

Figure 3.2 *The effects of atmospheric absorption on the spectrum of the sun*

35

If we consider the extreme wavelengths for blue and red light, of 400 nm and 700 nm respectively, we can calculate the ratio of the intensity of scattered blue light to the intensity of scattered red light. That is

$$\frac{\text{Intensity (blue)}}{\text{Intensity (red)}} = \frac{\lambda^4 \text{ (red)}}{\lambda^4 \text{ (blue)}} = \frac{(700 \times 10^{-9})^4}{(400 \times 10^{-9})^4} \approx 10$$

We see that when light is scattered by the particles of the atmosphere, the intensity of scattered blue light is ten times the intensity of scattered red light. This accounts for the blue colouration of the sky.

In addition, the intensity of scattered light is proportional to the square of the volume of the particle, so that scattering by atoms and molecules is slight. However, the earth's atmosphere contains such a number of molecules that the total intensity of the scattered light is easily observable. Without this scattering effect the sky would appear completely black, except in the direction of the sun.

The red colour of the rising or setting sun can also be explained in terms of the scattering of light by small particles. Figure 3.3 illustrates how light from the sun on the 'horizon' traverses an increased distance through the atmosphere to an observer.

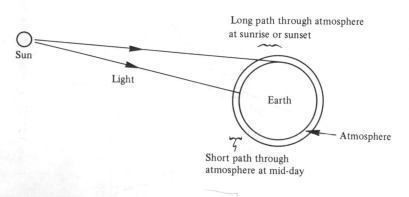

Figure 3.3 *Showing the relative distances travelled by sunlight through the earth's atmosphere*

Under these circumstances the blue light is scattered out before it reaches the observer and the sun appears red.

In the case of 'larger' dust particles the Rayleigh fourth power law does not hold, and the intensity of scattered light is independent of wavelength. Since they are heavier, larger dust particles are found in the lower atmospheric layers, and since at sunrise or sunset the sun's rays, already depleted in blue light in the upper atmosphere, travel a long path through the lower atmospheric layers where the remaining red light is scattered; this explains why the sky itself takes on a redder appearance.

The scattering of light by a particle which is large compared with the wavelength of the light is complex, and involves the processes of diffraction, diffuse reflection and refraction. As already stated, the scattering of light from a cloud of such particles is independent of wavelength, and so a shaft of sunlight entering a dark room may become visible because of the white light scattered by dust particles in the air.

Infrared Wavelengths

Infrared wavelengths are sufficiently large not to be diffracted by small dust particles, and so infrared radiation is able to penetrate dust clouds, where visible radiation is obscured. The importance of this is discussed in chapter 11 where we consider the penetration of interstellar dust clouds by infrared radiation.

Infrared radiation is however, partially absorbed by molecules of water and carbon dioxide in the earth's atmosphere, and so infrared observations from space must be carried out from satellites orbiting above the atmosphere. Also if the dust particles are very large compared with infrared wavelengths, then the light is absorbed rather than scattered. For example, infrared radiation will penetrate a fine mist but is attenuated in thick fog.

The Effect of the Altitude of the Sun on the Extent of Absorption

The effects of atmospheric absorption are less if the sun, or any other star, is observed when it is directly overhead, since radiation from a star at high elevation passes through less atmosphere to reach the earth's surface. Those absorption lines which are due to absorption in the earth's atmosphere show up to a much greater extent when the star is observed at low elevation. This is illustrated in figure 3.4.

ABSORPTION BY THE SUN'S 'ATMOSPHERE'

The sun is entirely gaseous and has no solid or liquid surface. However, the sun is stratified into many different levels, which show characteristic features and types of activity. A schematic diagram of the structure of the sun is shown in figure 3.5.

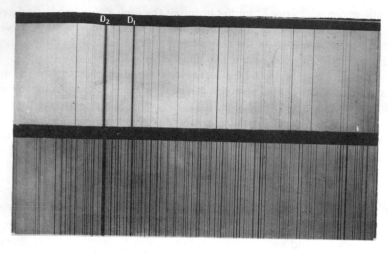

Figure 3.4 *High and Low Sun. 3 metre grating spectrograms, showing absorption lines. 1: High Sun; 2: Low Sun – by Higgs, Liverpool (photograph reproduced by permission of The Royal Astronomical Society)*

The innermost region or 'core' is the site of the thermonuclear reactions which generate the sun's energy. Most atoms in the core are broken down into sub-atomic particles, which move with very high velocities. Under these conditions we describe matter as being in the 'plasma' state. The pressure of the plasma is 10^9 atmospheres and it has a temperature of 1.4×10^7 K.

It is the fusion of protons within the plasma which is responsible for the release of energy. Photons radiating from this region are very energetic and are predominantly gamma or X-ray photons. The thermonuclear reactions within the core occur in a 'controlled' manner, since for a reaction to occur the pressure and temperature of the plasma must be sufficiently high. There is an equilibrium between gravitational forces which compress the sun and increase its temperature, and the resulting increase in pressure which resists further collapse. Radiation pressure tends to expand the core. If the rate of energy production increases, then the core expands, with the result that the temperature and pressure fall. Energy production then decreases and the core contracts, and so thermal equilibrium is maintained.

No part of the core is visible from outside the sun, and the deepest region which is observed is marked by the lower surface of the next region known as the 'photosphere'. The gases in this region are transparent. The photosphere has a low pressure of 10^{-2} atmospheres and a temperature of about 6000 K (see chapter 2).

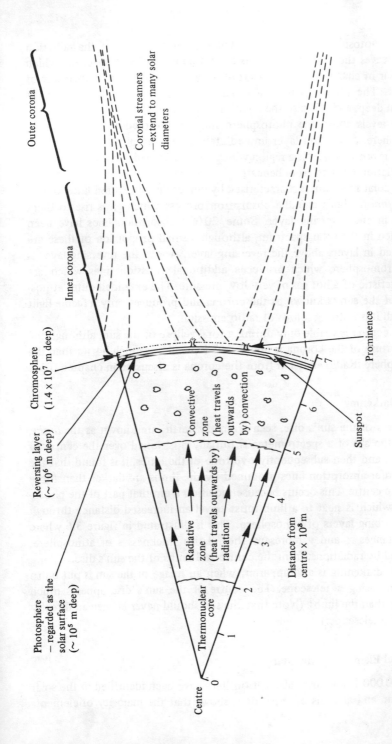

Figure 3.5 *A section through the sun*

Outer corona

Coronal streamers
—extend to many solar
diameters

Inner corona

Chromosphere
(1.4×10^7 m deep)

Prominence

Reversing layer
($\sim 10^6$ m deep)

Convective
zone
(heat travels
outwards
by) convection

Sunspot

Photosphere
—regarded as the
solar surface
($\sim 10^5$ m deep)

Radiative zone
(heat travels outwards by)
radiation

Thermonuclear
core

Centre

Distance from
centre $\times 10^8$ m

0   1   2   3   4   5   6   7

39

The photosphere is the region which produces most of the radiation which leaves the sun, and it yields a continuous emission spectrum which is similar in character to the spectrum of a black body at a temperature of 6000 K. The photosphere is a relatively thin layer, being some 100 or 200 km deep, and it is also the region in which 'sunspots' are observed.

The levels above the photosphere may be regarded as being the solar atmosphere, and a thin layer immediately above the photosphere, known as the 'reversing layer' is a region where there is considerable absorption of the radiation released from beneath.

The solar spectrum is characterised by numerous absorption lines known as *Fraunhofer* lines and the absorption process which gives rise to them occurs in the reversing layer. Some 20 000 absorption lines have been identified in the solar spectrum, although a small proportion of these are produced in layers above the reversing layer. Above the reversing layer is the chromosphere which produces additional emission lines which are characteristic of a hot gas of very low pressure. The extended outer atmosphere of the sun is known as the corona and produces only a feeble light although it is a strong source of radio emission.

The corona is visible only during a total eclipse of the sun, although the temperature of the corona is about 10^6 K, which is much hotter than the photosphere. Radio emission from the corona is discussed in chapter 10.

Limb Darkening

The edges of the sun's disc, seen from the earth, are known as the *limbs*, and if the slit of a spectrometer is initially positioned over the centre of the disc, and then subsequently over one of the limbs, it is found that the Fraunhofer absorption lines produced by the limbs are darker than those from the centre. This occurs because radiation from that part of the photosphere which is near to a limb must travel an increased distance through the absorbing layers of atmosphere. This is illustrated in figure 3.6 where the distances x and y represent the relative thicknesses of atmosphere, traversed by radiations, from the centre and limbs of the sun's disc.

Limb darkening is also apparent when an image of the sun is put on to a screen, using a telescope. The centre of the sun's disc appears much brighter than the limbs. (Note that the sun should never be viewed directly through a telescope.)

Chemical Elements in the Stars

Some 20 000 Fraunhofer absorption lines have been identified in the solar spectrum, and analysis of these lines shows that the majority of elements

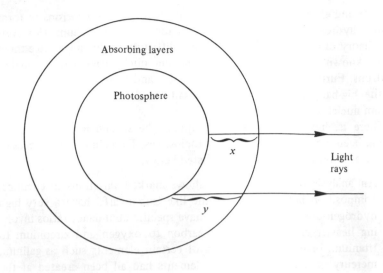

Figure 3.6 *Illustrating 'limb darkening'*

arc present in the sun's atmospheric layers, although in most cases they occur in relatively minute quantities. Information regarding the relative abundances of the elements can be gleaned by examining the 'relative' line strengths. It is found that hydrogen is the most abundant element, there being 20 times more hydrogen than helium, and the other elements account for about 1/1000 of the total quantity of the matter. It should be noted that in the case of the sun, temperatures are too low to produce absorption lines for helium, and its abundance has to be calculated from estimates based upon its contribution to the total gas pressure.

The question of where the elements were formed is a perplexing one. All stars exhibit the presence of elements in their spectra, and in chapter 9 we see how this leads to an important method of stellar classification.

A number of theories have been proposed to explain the creation of the elements but all are essentially variants of two basic ideas which are that the elements were created at the time when the universe itself was created, or that hydrogen was the original element and all other elements were created from it by nuclear reactions within stars.

The first of these ideas is associated with the 'big-bang' theory of the origin of the universe, which assumes that the entire matter of the universe was at one time contained together and compressed into a substance called 'ylem'. The ylem would have such an incredibly high density and temperature that the matter consisted essentially only of neutrons, but then as the universe started to expand, the neutrons broke down into protons and electrons. The protons would then capture neutrons to form 'heavy' hydrogen, which would then combine to form helium. However, this theory of element-building runs into difficulty at this stage since there is no known mechanism by which helium nuclei can capture further neutrons. Further, the conditions of density and temperature at this stage of the big-bang would not be sufficient for the direct combination of helium nuclei to form heavier elements.

There is, however, evidence to support the second idea of elements having been created within the stars themselves. Three important pieces of evidence supporting this theory are listed below.

(i) An analysis of stellar spectra shows marked variations in chemical composition between stars: some like 'sub-dwarfs' have a very high hydrogen-to-metal ratio; others have specific abundance ratios involving heavier elements, such as carbon to oxygen and zirconium to titanium, or excessive amounts of certain elements such as gallium, mercury or manganese. If the elements had all been created at the same time, then it would be very difficult to reconcile this very unusual distribution.

(ii) Stars of high hydrogen-to-metal ratios are 'old' stars, which indicates that as a result of their slow evolution they have not yet reached the nuclear burning stages which yield the heavier elements. Younger stars are made of 'recycled material' obtained from stars which have had occur within them a whole series of nuclear reactions, in which heavier elements were synthesised, before the stars became extinct and returned their matter to the interstellar medium.

(iii) The relative abundances of elements within the solar system cannot be explained in terms of a single process or set of conditions, which again implies the elements were not created at one time.

These three pieces of evidence suggest that all elements, with the exception of hydrogen, have been manufactured in stars, and furthermore that stars exhibit an evolutionary sequence, which makes possible the production of all elements by various nuclear reactions, starting with hydrogen (see chapter 9). To produce the elemental abundances which are observed in the stars requires that quite a number of distinct types of nuclear reaction have occurred. Among these are included the proton–proton chain, the carbon cycle and the triple-alpha process, which are relatively straightforward reactions. More complex element-producing processes occur under special conditions, such as are obtained when an unstable star explodes as a 'supernova'.

The material from which the sun and solar system is composed seems to have had a complicated history. Some of this material, at least, must have been processed through two or perhaps more generations of stars, and the naturally occurring radioactive elements such as uranium and thorium may have been 'seeded' by radioactive elements produced in the explosion of a supernova.

4 The Eye as a Detector of Electromagnetic Radiation

Our 'visual perception' depends not only upon the characteristics of the electromagnetic radiations emitted by a source, but also upon the physiology of the 'eye–brain' system. The images that we obtain in our minds are the result of photochemical–neural processes which occur in the retina of the eye, and also the psychological responses which are made by the brain. An example of a visual effect which is due to the physiology of the retina, is that objects appear to lose their colour in dim light. This 'apparent' loss of colour is due to the fact that in bright conditions, 'cone' cells in the retina are 'triggered' to respond, whereas in dim light 'rod' cells, which are especially adapted and interconnected for gathering diffuse light, are the agents involved in producing the visual response. It is only when cone cells are involved that we perceive colour.

When a faint celestial object is observed through an optical telescope, the eye is operating in 'dim' light conditions and will not perceive colour. For example, when faint nebulae are observed by the eye through a telescope the images which are perceived are nearly always in monochrome. It is interesting to reflect that the colour sensation itself is purely a figment of our minds. If we look at a red or green object for example, then the red or green colour exists only in our minds. The two objects are simply emitters of electromagnetic waves which are distinguished only by having different frequencies. Furthermore, colours are not produced in our eyes since the optic nerve which leads from the eye to the brain carries nothing but electrical impulses. However, these electrical impulses do have encoded in them the information that the electromagnetic radiations being emitted by the two objects have different frequencies. It is then a psychological response within our minds which creates the colour sensations of red and green.

To understand these and other aspects of our visual perception which are relevant to astronomy, we need to know something of the structure and functioning of the eye.

THE STRUCTURE AND FUNCTIONING OF THE EYE

A sectional view of the eye is shown in figure 4.1.

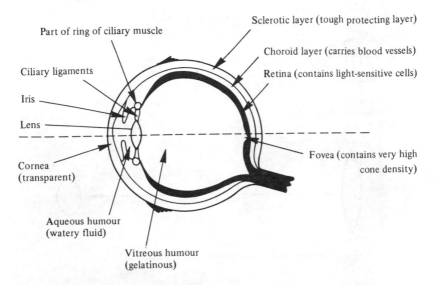

Figure 4.1 *Section through the human eye*

An inverted image is formed on the retina by successive refraction at the surfaces between the air, the cornea, the aqueous humour, the lens, and the vitreous humour. The cornea has the greatest effect in converging incoming rays of light, but the final focusing is obtained by the curvature of the lens being altered. This process is known as *accommodation*.

The normal eye has a 'far point' at infinity, which means that there is no limit to the distance we can see, provided the source is bright enough.

The 'near point' or 'least distance of distinct vision' is at about 25 centimetres from the eye and is the shortest distance at which an object can be viewed in focus. When viewing distant objects the ciliary muscle, which controls the shape of the lens, is relaxed and the lens becomes thinner. Conversely when viewing close objects, the ciliary muscle contracts and allows the lens to become more rounded. The ciliary muscle is in the form of a ring and is connected to the lens by ciliary ligaments. These appear as fine black threads in a dissected eye. When the ciliary muscle contracts it tightens and makes a smaller ring. The ciliary ligaments have less tension and the lens assumes a more rounded shape. When the ciliary muscle relaxes, a large ring is formed and tension is created in the ciliary ligaments and the lens becomes thinner. This is illustrated in figure 4.2.

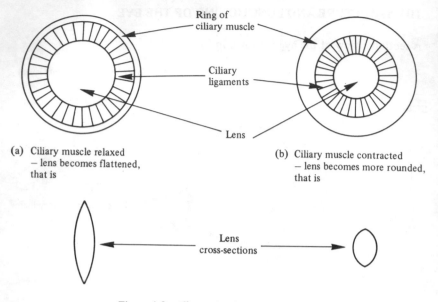

(a) Ciliary muscle relaxed
— lens becomes flattened,
that is

(b) Ciliary muscle contracted
— lens becomes more rounded,
that is

Figure 4.2   *Illustrating 'accommodation'*

The eye controls the amount of light entering by a process known as *adaption*. The iris or coloured part of the eye increases or decreases in area to control the size of the pupil. This is shown in figure 4.3.

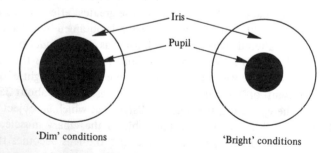

'Dim' conditions   'Bright' conditions

Figure 4.3   *Illustrating 'adaption'*

The eye can adapt to a range of brightness of 1:1 million, but the range of movement of pupil diameter is only four-fold, and so the control of pupil size is not the only compensatory mechanism for brightness. The retina itself is responsible for much of the adaption. A schematic diagram of a cross-section through the retina is shown in figure 4.4.

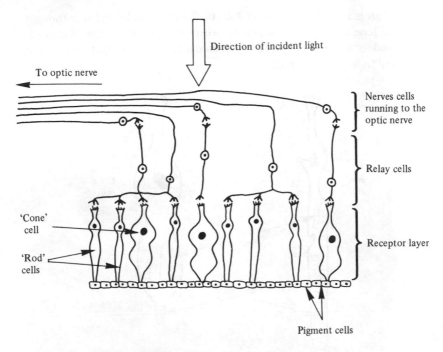

Direction of incident light

To optic nerve

Nerves cells running to the optic nerve

Relay cells

'Cone' cell

Receptor layer

'Rod' cells

Pigment cells

Figure 4.4 *Schematic cross-section through the retina*

Note: There are many other cells present in the retina which inter-connect the light-sensitive cells, but these are omitted from this diagram for the sake of clarity!

THE SHAPE OF THE RECEPTOR CELLS

At the back of the retina is a single layer of cells which contain the black pigment melanin. Directly above this layer are the rod and cone cells, and the structure of these two types of cell is shown in more detail in figure 4.5.

The rods are single cells which are extended at their outer ends to form long slender cylinders. These cylinders contain a purple photosensitive pigment known as rhodopsin, or visual purple, which is closely allied to Vitamin A. The cones are also single cells but are 'fatter' and their outer ends form small tapering processes; which contain other pigments.

It is thought that the 'bulging' cone may be so shaped as to converge rays of light, which are travelling parallel to the central axis of the eye, on to the small tapering end portion which contains the photosensitive pig-

47

ment. This is illustrated in figure 4.6. In the case of the rod it is thought that the long tubular outer segment, where the photosensitive pigment is contained, is adapted for capturing minimal amounts of light falling on it diffusely from different angles.

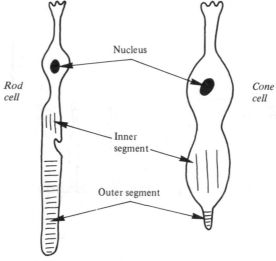

The photosensitive pigment is arranged in
horizontal bands in the outer segments

Figure 4.5   *The structure of a rod cell and a cone cell*

THE DISTRIBUTION OF RODS AND CONES IN THE RETINA

Rods and cones are not evenly distributed over the retina. The *fovea centralis* is a small depression in the retina (see figure 4.1) where only cone cells are found, and in very high density. Away from this area a mixture of rods and cones is found, and towards the periphery of the retina the cones are greatly outnumbered by the rods.

Cones do not operate in dim light conditions, and an observer will be more likely to detect a faint star if the eye's gaze is directed slightly to the side of the star. In this case the image of the star forms on the rod cells at the periphery of the retina, rather than on the cone cells at the fovea. The most sensitive region is some 10 to 15° 'off' the fovea.

The rods and cones are connected via relay cells to nerve cells which form the optic nerve.

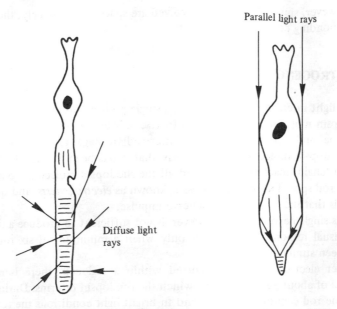

Parallel light rays

Diffuse light rays

Figure 4.6 *Light capture in a rod and a cone*

The *resolving power* of the eye depends upon the optical system and also upon the closeness of the receptor cells in the retina. The fovea represents the area of densest packing and contains about 150 000 cones per square millimetre. There are no blood vessels in this spot and the nerve fibres above the cones diverge, so that light is admitted directly. Over other areas of the retina light must first find its way through the layers of nerve fibres before it impinges upon the receptor cells. In good light the eye automatically focuses the centre of the image on the fovea, which can resolve two points of separation 1/10 millimetre at a distance of 25 cm from the eye. Other areas of the retina are much less discerning.

The human retina has been calculated to contain over 100 million rods, 6 million cones and about 1 million optic nerve fibres. Recent studies have shown that no cells, rods or cones share a 1:1 link with an optic nerve cell, but instead the rods and cones share inter-connections which are highly elaborate, although inter-connections among the rods are the more developed. It is these inter-connections between rod cells which enables them to operate in twilight conditions, where the summation of the responses of a number of rods within a group becomes significant. In these conditions minimal amounts of light falling on many rods can give a detectable visual response.

However, since the rod cells involved are spaced more widely, there is a corresponding decrease in resolution.

ELECTROGENESIS

When light interacts with rhodopsin, a single photon interacts with a single rhodopsin molecule, and splits it into two substances which are known as 'retinene' and 'opsin'. A number of intermediate stages are involved. When the rhodopsin molecule is split we say that it has been 'bleached'. At this stage a 'chain' reaction follows and all the rhodopsin molecules contained in the rod split. This whole process is known as *electrogenesis*, and the end result is that the rod cell 'fires' a nerve impulse.

This single nerve impulse, however, is not sufficient to produce a detectable visual response. This occurs only when a minimum of six rod cells have been stimulated.

After electrogenesis has occurred within a rod cell, there is a time interval of about a second during which the rhodopsin reforms. During this time the rod cannot 'fire' again and in bright light conditions the rods are permanently bleached and are inoperative. The recovery of the rod cells may be observed when a room is suddenly plunged into darkness. At first it is impossible to see anything, but then the rod cells slowly become operative under the conditions of twilight vision.

If a photographic plate is left exposed to the stars, then the effect of impinging photons is cumulative, so that the development process proceeds as long as the plate is left exposed. Consequently, good photographs may be taken of very faint stars by extending the exposure time. It should be noted that there is no such cumulative effect in the eye. A single photon can cause electrogenesis in a rod cell, but no matter how many photons arrive immediately afterwards the strength of the signal, put out by the cell, is no greater. Neither will prolonged exposure of the retina make a faint star more visible, for the effects of electrogenesis are immediate, and there is no storage mechanism to enhance a weak signal.

THE QUANTUM EFFICIENCY OF THE EYE

It has been determined experimentally that for a point source of light to be detectable, the minimum energy rate for light striking the eye must be 10^{-16} Watts.

Note that a 'point' source of light produces a 'point' image, and so we are considering light falling upon a very small area of the retina.

If we consider light of wavelength, say, 500 nm, then the corresponding energy of a photon is

$$\text{Energy of photon} = hf = \frac{hc}{\lambda} = \frac{6.62 \times 10^{-34} \times 3 \times 10^8}{500 \times 10^{-9}} \approx 4 \times 10^{-19} \text{ J}$$

and the number of photons arriving at the eye each second is

$$\frac{10^{-16}}{4 \times 10^{-19}} = 250$$

These figures indicate that some 250 photons must arrive at the eye each second, for a point source to be just visible.

Many of these original photons do not arrive at the retina and are 'wasted'. Some 2 per cent of the light incident on the cornea is lost by reflection, and for light of wavelength 500 nm, 50 per cent of the remaining photons are lost by absorption and scattering within the various media on the way to the retina. We are left with a figure of 120 photons which reach the retina.

An experiment has shown that rhodopsin itself absorbs only 20 per cent of the light which is incident upon it. In this experiment the percentage of light reflected back from a 'dark adapted' retina was compared with the percentage of light reflected back from a retina illuminated by a bright light, in which case all the rhodopsin molecules would be 'bleached' and unable to accept any further photons. The figure of 20 per cent indicates that from the 120 photons incident on the retina, only about 24 are absorbed by the pigment rhodopsin.

The quantum efficiency of the eye is measured as the ratio of the number of photons which react with a rhodopsin molecule and cause 'electrogenesis', to the numbers of photons originally incident on the cornea. The quantum efficiency of the eye is therefore

$$\text{Quantum efficiency} = \frac{24 \text{ photons reacting with rhodopsin}}{250 \text{ photons arriving at cornea}} \approx 10 \text{ per cent}$$

Thus the quantum efficiency of the eye is approximately 10 per cent.

THE SPECTRAL SENSITIVITY OF THE EYE

Twilight vision differs from bright light vision not only in the presence or absence of colour, but also in that the spectral sensitivity alters according to whether rods or cones are in operation. The rods have their peak sensitivity at a wavelength of 510 nanometres (blue-green), whereas the cones peak at 560 nanometres (yellow-green). Curves showing the spectral sensitivities of rods and cones in the human eye are given in figure 4.7. It can be seen from these graphs that the cones respond to a deep red light where the rods will detect nothing. Thus a blue and a red object, in which the red may appear brighter in bright light, will in the dark be in a reverse position with the blue object appearing brighter. The shift in relative brightness of coloured objects as the intensity drops below the cone threshold or rises above it, is called the 'Purkinje phenomenon'. It is easy to observe if one opens a colour magazine in the twilight.

After studying the brightest areas (in shades of grey), these areas are found not to be the brightest when the magazine is illuminated by bright light.

The spectral sensitivity curve for cones shown in figure 4.7 is in fact a composite curve, made by adding the spectral sensitivity curves for three different groups of cones which have been identified in the human retina. The first group of cones responds to the 'blue' light in the region of 400 to 500 nm, the second group responds to 'green' light in the region of 450 to 630 nm, and the third group responds to 'red' light in the region of 500 to 700 nm. The spectral sensitivity curves for each type of receptor are shown in figure 4.8. By adding the ordinates of these three curves, we obtain the spectral sensitivity curve for 'bright light' conditions for the eye as a whole. Hence the 'resultant' curve in figure 4.8 is the same as the curve for cones shown in figure 4.7.

The resultant curve shows that the retina does not respond equally to all wavelengths. Blues, for example, when compared with shades of orange or green, appear dark, whereas yellows, including the deepest shades, appear bright. The extent to which the visual response for the brightest colour, yellow-green, exceeds that for colours nearer the ends of the 'visible' spectrum is quite pronounced.

THE EYE'S RESPONSE TO INTENSITY

The eye's response to intensity is logarithmic. This means that the eye's response to brightness is equal to a constant, multiplied by the natural logarithm of the actual change in intensity. The eye's response is markedly small for any given change in actual intensity.

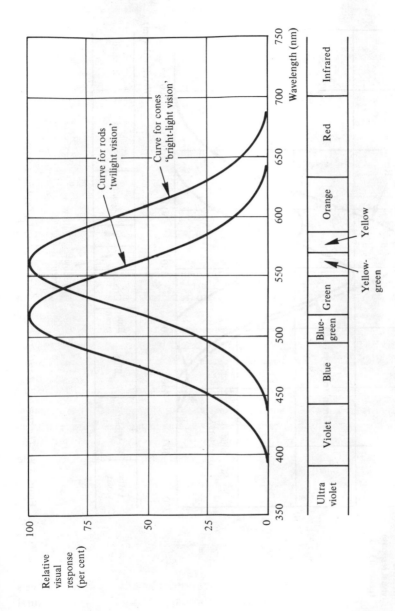

Figure 4.7 The spectral sensitivity of the eye

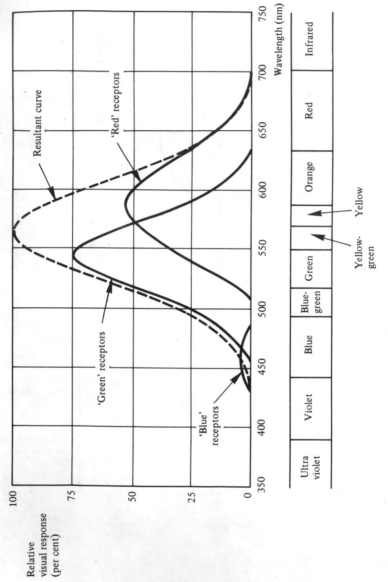

Figure 4.8 *The spectral sensitivity curves for three groups of colour receptor cone cells*

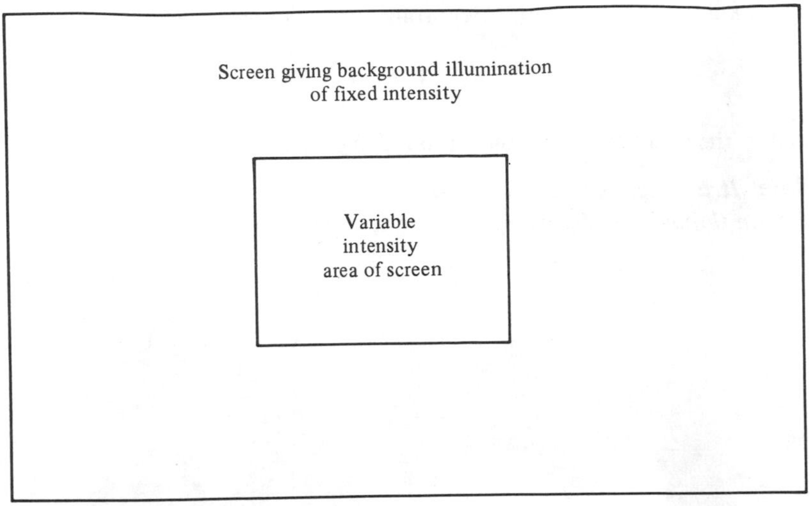

Figure 4.9 *The investigation of the eye's response to intensity of illumination*

The logarithmic response of the eye may be demonstrated by experiments in which the intensity of an observed light is varied, in relation to a background light of fixed intensity, as illustrated in figure 4.9.

If the two areas of the screen are initially the same brightness, then the inner area can be increased in intensity by an amount δI which then makes the inner area just detectably different. The results show that for a series of background intensities I

$$\frac{\delta I}{I} = \text{constant } (c)$$

This is known as the *Weber–Ffechner* Law.

If we say that δS is the 'small sensation response' of the eye at which the brain just perceives a difference in sensation for the intensity change δI, against a background of intensity I, then we can say that

$$\delta S \propto \frac{\delta I}{I}$$

whence we have

$$\delta S = k \frac{\delta I}{I}$$

where k is another constant. Integrating both sides we have

$$S = k \log_e I$$

that is, the measured sensation is related logarithmically to intensity.

Note: *It is dangerous to observe the sun visually and an attempt to view the sun through a telescope may result in damage to the eye.*

5 The Photographic Process as a Means of Detecting Electromagnetic Radiation

After considering the limitations of our visual capacities, we now look at the uses of photography in astronomy. We begin by looking at the 'silver bromide' photographic process.

THE SILVER BROMIDE PHOTOGRAPHIC PROCESS

This is the most common system used for photography, and involves the photochemical transformation of silver bromide (AgBr) to silver (Ag). A photographic film consists of a plastic transparent backing, coated with the photosensitive emulsion. The emulsion consists of finely divided silver bromide, suspended in a solution of gelatine. A variety of other compounds may be added to the emulsion to obtain specific sensitivities to intensity or wavelength, but the basic photochemical process is still the transformation of the silver bromide. The structure of a photographic film is shown in figure 5.1.

Exposure of the film to light reduces the silver bromide to silver, and by exposing the film for a long enough time period, it is possible to pro-

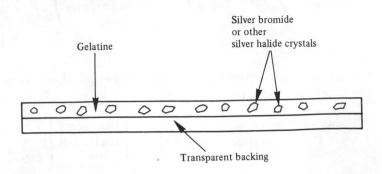

Figure 5.1   *The photographic film*

duce an image on the film. Usually, however, the film is given only a short exposure and an invisible 'latent' image of silver is formed.

The film is then 'developed'; it is treated with a reducing agent such as hydroquinine (Quinol) or iron(II) oxalate, which converts more of the silver bromide to silver. What is most important is that the latent image of silver acts as a catalyst for the further reduction of silver bromide, so that during development a silver image forms where the latent image was present. Development thus acts as an 'amplifying' process.

Finally, the film is 'fixed' — the silver bromide remaining is dissolved away by a solution of sodium thiosulphate for example; that is

$$AgBr_{(s)} + 2S_2 O_3^{2-} \rightarrow [Ag(S_2 O_3)_2]^{3-} + Br^-_{(aq)}$$

This leaves an image of silver, deposited in proportion to the amount of light received by that part of the film. It should be noted that because of their small size, the grains of silver appear 'black'. A summary diagram of the photographic process is shown in figure 5.2.

Later we compare the efficiency of the retina with the efficiency of the photographic plate, and for this purpose we will examine the chemistry of the silver bromide process in detail.

In the following account of the chemistry of the silver bromide photographic process, the numbers of photons which have been used are totalled up in the 'rectangular boxes'.

In detail then, the initial photo-reaction produces one electron when a photon falls on to a bromine ion in the silver bromide grain

$$Br^-\ +\ (hf)\ \rightarrow\ Br^{\bullet}\ +\ e^-$$

$$\begin{pmatrix} \text{bromine ion} \\ \text{with one unpaired} \\ \text{electron} \end{pmatrix} + \text{(photon)} \rightarrow \begin{pmatrix} \text{bromine radical} - \text{has} \\ \text{equal numbers of} \\ \text{positive and negative} \\ \text{charges} \end{pmatrix} + \text{(electron)}$$

$$(1)$$

This electron may recombine with the bromine radical, otherwise it may be trapped to form a silver atom

$$Ag^+ + e^- \longrightarrow Ag$$
$$\text{(pre-image speck)}$$

| one photon used at this stage | (2) |

This atom of silver is known as the 'pre-image speck', and it has a very short life.

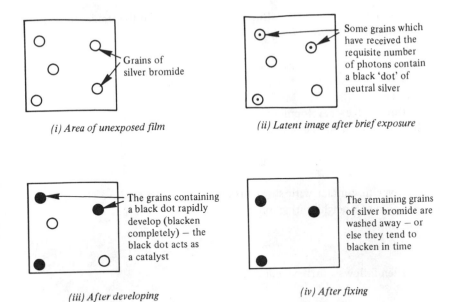

(i) Area of unexposed film

Grains of silver bromide

Some grains which have received the requisite number of photons contain a black 'dot' of neutral silver

(ii) Latent image after brief exposure

The grains containing a black dot rapidly develop (blacken completely) – the black dot acts as a catalyst

(iii) After developing

The remaining grains of silver bromide are washed away – or else they tend to blacken in time

(iv) After fixing

Figure 5.2   *Summary diagram of the silver bromide photographic process*

The next stage occurs when a second electron, produced as in equation (1), arrives at or near the pre-image speck, and forms a second silver atom which combines with the first to give the 'sub-image' – Ag_2

$$Ag + Ag^+ + e^- \longrightarrow Ag_2 \qquad \boxed{\text{two photons used by this stage}} \qquad (3)$$
$$\text{(sub-image)}$$

The sub-image is more stable, but at room temperature it can decompose after several days. It is, however, difficult to develop.

The arrival of a third electron, again produced as in equation (1), initiates a further stage

$$Ag_2 + Ag^+ \rightleftharpoons Ag_3^+ \qquad (4)$$
$$\text{(an unstable species)}$$

then

$$Ag_3^+ + e^- \longrightarrow Ag_3 \qquad \boxed{\text{three photons used}} \qquad (5)$$

59

Ag_3 readily combines with a further ion to form the stable, tetrahedral Ag_4^+ which is the *latent image*

$$Ag_3 + Ag^+ \longrightarrow Ag_4^+$$
$$\text{(latent image)} \tag{6}$$

During the 'development process', the developing chemical supplies electrons, allowing the latent image to grow

$$Ag_4^+ + e^- \longrightarrow Ag_4 \tag{7}$$

Silver in contact with silver bromide acquires a positive charge by the inclusion of a neighbouring ion

$$Ag_4 + Ag^+ \longrightarrow Ag_5^+ \tag{8}$$

Then follows a further reduction

$$Ag_5^+ + e^- \longrightarrow Ag_5 \tag{9 etc.}$$

and so the process continues with the small speck of silver growing until the whole grain is blackened.

The development process is an 'amplifying' process, since a single grain may contain up to 10^{10} silver atoms which develop from the original four atoms made available by photon capture.

THE QUANTUM EFFICIENCY OF A PHOTOGRAPHIC EMULSION

From the foregoing account of the chemistry of the photographic process, we see that a minimum of three photons are required to produce the latent image, and so render a grain developable. The figure of three photons may be higher for other types of silver halide, and in some instances as many as ten photons may be required. One blackened grain on a photograph is sufficient to show the presence of a star. However, photographic emulsions have very low quantum efficiencies and far more than the minimum number of photons are required for latent image formation. The 'highest' value for the quantum efficiency of a photographic emulsion is about 2 per cent.

The reasons for the wastage of photons are as follows

(i) Some 90 per cent of the light which falls upon the emulsion is either diffusely reflected or diffusely transmitted. Only 10 per cent of the incident light is absorbed by the grains.

(ii) Some of the photons which are absorbed in the grains are wasted, if the grain they enter has already acquired enough photons for the formation of the latent image.

(iii) The reactions involved in the formation of the latent image are reversible, and unless enough photons arrive in a given time to form the 'stable' latent image, then earlier reactions may reverse and the photons involved in those will have been wasted.

A Formula for the Quantum Efficiency of a Photographic Emulsion

The quantum efficiency (Q.E.) of an emulsion may be defined by

$$Q.E. = \frac{\text{the number of photons arriving in unit time upon the exposed surface of grain } (n')}{\text{the number of silver atoms present at any one time within the grain } (n)}$$

Let N be the number of photons falling upon unit area in unit time, and let the area which the grain presents to the flux of photons be A, then $n' = NA$, and

$$Q.E. = \frac{n'}{n} = \frac{NA}{n} \tag{i}$$

The 'Intensity' (I) of radiation is defined as the rate of flow of energy across unit area. The units of intensity are Watts m^{-2}.

If N photons cross unit area in time t, and each photon carries energy hf, then the intensity of radiation is

$$I = \frac{N(hf)}{t} \quad \text{and} \quad N = \frac{It}{(hf)}$$

Substituting for N in equation (i), we have

$$Q.E. = \frac{It\,A}{(hf)n} \tag{ii}$$

The 'Exposure' (E) to which a surface is subjected is defined as the product of the intensity of the radiation and the 'exposure time', that is

$$E = It$$

The units of exposure are Joules m^{-2}.

So in terms of exposure, equation (ii) becomes

$$Q.E. = \frac{EA}{(hf)\,n} \qquad \text{(iii)}$$

Thus if all the quantities apart from n in equation (iii) combine so that n becomes, say, 3 for silver bromide, then a latent image will be formed.

Equation (iii) shows that n is proportional to the exposure, so that more grains are darkened if the exposure is greater. This relationship is not true at very low or very high exposure levels. This is discussed in more detail when we examine the effect of exposure on image density.

Equation (iii) also shows that the quantum efficiency is greater when the exposed area of grain is larger, and also that the quantum efficiency is inversely proportional to the frequency of radiation.

THE CUMULATIVE EFFECT OF LIGHT ON A PHOTOGRAPHIC EMULSION

The degree to which a photographic film is blackened depends upon the amount of exposure of the film, rather than just the intensity of radiation. This is evident from equation (iii). It is possible for a small intensity and a large exposure time to have the same effect as a large intensity and short exposure time, provided the product It is the same in each case.

Thus a film can 'accumulate' photons over a longer period of time, and in this way very faint objects in space may be photographed. No similar effect exists either for the eye or the photoelectric cell, discussed in the next chapter. In the case of the eye, each photon that reacts with a rhodopsin molecule immediately causes electrogenesis. If other cells are not stimulated to respond at the same time, then the photon is wasted, since it requires electrogenesis in a number of cells to produce a visual response. The effects of the photon are immediate, and after electrogenesis the cell soon recovers its original unbleached state — there is no storage mechanism for photons in the eye!

With long exposure times the photographic film can detect objects in space far too faint to be seen with the eye.

GRAIN SIZE AND FILM SPEED

If the exposure area of a grain is large, it follows that it will acquire the number of photons required to produce the latent image in a shorter period of time. This is because the grain offers a larger collecting surface. Consequently a large grained film will be a 'fast' film and will blacken after a short exposure time.

The problem with large grain size is that it gives poor resolution, and small-sized grains are essential for astronomical work, where it is often wished to resolve stars which have a very small angular separation. Astronomers are, in any case, able to use long exposure times, since astronomical telescopes are often 'driven' to compensate for the rotation of the earth.

We see from equation (iii) that if the size of the grains is reduced, then the quantum efficiency drops, and 'fast' films which are used in astronomy as 'high resolution' films may have quantum efficiencies as low as 0.16 per cent. These very low quantum efficiencies are, however, very well compensated for by the high amplification factor which occurs in the development process.

IMAGE ENLARGEMENT DUE TO THE SCATTERING OF PHOTONS WITHIN THE EMULSION

The scattering process which we examined in chapter 3 also occurs within a photographic emulsion. In this case light is scattered by the grains themselves, and the result is that the photographic image covers a wider area then the original optical image which was formed on the film. However, the effect is unequal and the extent of scatter is greater for brighter objects. So, for example, if one star appears brighter than another visually, then the difference in brightness becomes more apparent on a photograph. The increase in the size of the image due to scattering is illustrated in figure 5.3.

'GREY SHADING'

The arrival of photons on the surface of the emulsion is a random process, and while the average number of effective photons per grain is given by

$$n = \text{Q.E.} \times NA$$

any individual grain may receive more or less than this average number.

If we consider a group of several hundred grains, which would appear as a small dot to the eye, then some of these grains will receive enough photons for the formation of the latent image, while others will not. After

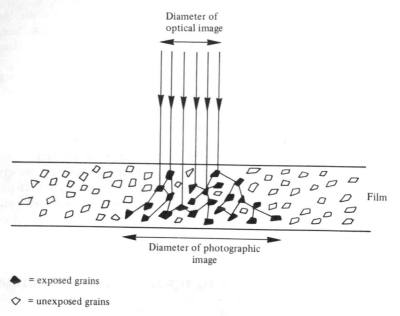

Figure 5.3 *Illustrating the effects of scattering on the size of the photographic image*

development the dot will appear as a shade of grey, and the darkness of the shade rises as the fraction of developed grains rises.

MEASURING THE AMOUNT OF BLACKENING

When a film has been developed and fixed, the density of the image produced may be examined in terms of the amount of light which can be transmitted through the film. In figure 5.4 I_i represents an amount of incident light falling on to a film and I_t represents the amount of light which is transmitted. The amount of blackening which has occurred may be expressed in terms of these quantities.

There are three ways of expressing the 'blackness' of a negative. The *transmission* of an area of the negative is defined as the ratio of transmitted light to incident light, that is

$$\text{Transmission } (T) = \frac{I_t}{I_i}$$

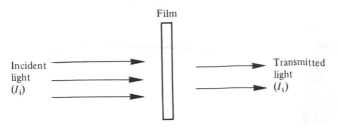

Figure 5.4   *Measuring the amount of blackening*

Transmission is always less than unity and is often expressed as a percentage.

The *opacity* is another expression, and is defined as the ratio of incident light to transmitted light, or

$$\text{Opacity } (O) = \frac{I_i}{I_t} = \frac{1}{T}$$

The unit of blackening which is most generally used is the *density*, and is defined as the natural logarithm of the opacity; that is

$$\text{Density } (D) = \log_e O = \log_e \left(\frac{1}{T}\right) = \log_e \left(\frac{I_i}{I_t}\right)$$

Density is the most preferred unit because it bears a simple numerical relationship to the amount of silver present, which is that the density doubles when the amount of silver doubles — that is, they vary in direct proportion.

As a result of the way density is defined, this term relates the degree of blackening to the way in which the eye assesses it, since the eye's response to changing light intensity is logarithmic.

If we examine various areas of a negative in which the density increases by fixed increments, then the eye accepts these successive increments as representing equal increases in the amount of blackening.

THE EFFECT OF EXPOSURE ON IMAGE DENSITY

Photographic emulsions of different types exhibit particular characteristics in their responses to exposure. For example, for a given exposure one film may show a greater image density than another.

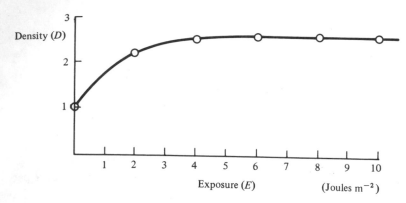

Figure 5.5 *A characteristic curve*

In general, a characteristic curve is obtained when density is plotted against exposure, and the general shape of such a curve is shown in figure 5.5.

Although a curve of this type is sometimes of value, a more useful curve is obtained by plotting density against the logarithm of exposure. This is known as a *D log E* curve, or an *H and D* curve after Hurter and Driffield who first published such curves. The characteristic curve shows the effect on the emulsion, of every degree of exposure from under-exposure to over-exposure, for a given development time and a given developing chemical. A typical $D \log E$ curve is shown in figure 5.6.

The *D log E* curve has an advantage over the *D versus E* curve, in that the portion of the curve which shows the region of 'just perceptible' blackening, appears on a larger scale. This region is known as the 'dynamic range of input' and is shown in figure 5.6. The speed of a film is usually judged in terms of the exposure needed to produce small changes in density rather than complete blackening, and so the enhanced clarity of this part of the curve is very useful.

Over the dynamic range of input, if the performance of the emulsion is such that a steep gradient is obtained, then a small range of exposure only is necessary to make the film go from transparent to black. The gradient at this section is known as γ and it provides a measurement of 'contrast' within the developed film.

Within the dynamic range of input is a portion of the curve which is a straight line. It is only on this straight-line region of the curve that density variations in the negative correspond proportionally to visual differences in the original scene. The straight-line region is therefore a region of 'correct' exposure.

66

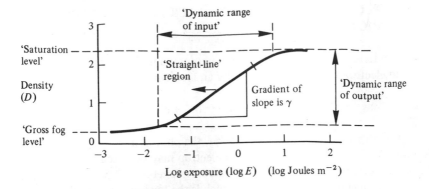

Figure 5.6 *A typical* D *log* E *curve*

Also shown in figure 5.6 are the 'gross fog level' and the 'saturation level'. The 'fog level' indicates that even if the film was exposed in total darkness, it would show a dull 'fog' since a proportion of silver (blackening) is produced by the developer. The 'saturation level' is reached when every grain has been blackened.

Finally, the 'film speed' may be deduced from the position of the graph on the scale. In figure 5.7, three D log E curves are shown on the same scale. The curve on the left is the fastest film, since it achieves any given density after the shortest exposure.

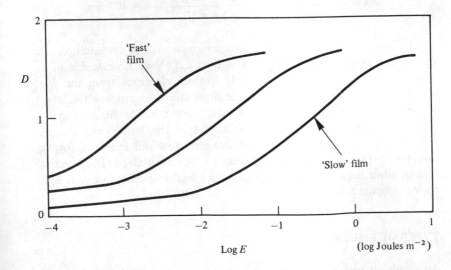

Figure 5.7 D *log* E *curves for films of different speed*

THE SPECTRAL SENSITIVITY OF 'BLACK AND WHITE' PHOTO-GRAPHIC FILMS

The initial step in the silver bromide photographic process is the liberation of an electron from a bromide ion by an incident photon

$$Br^- + (hf) \rightarrow Br^\bullet + e^-$$

For this reaction to occur, however, the photon must possess an energy in excess of 2.5 eV, in order that an electron may be raised from the upper energy level of the ion and be liberated. An energy of 2.5 eV is equivalent to a wavelength of about 497 nm which is the longest wavelength to which a silver bromide film will respond. This wavelength is within the blue-green region, and such a film will not be developed by any green, yellow, orange or red light. We calculate the wavelength from the energy required to liberate an electron as follows

$$\text{Energy of the photon in joules} = hf = \frac{hc}{\lambda}$$

To convert the energy in electron volts to Joules, we multiply by the electronic charge of 1.6×10^{-19} C, whereupon

$$\lambda = \frac{hc}{E_{(\text{Joules})}} = \frac{6.62 \times 10^{-34} \times 3 \times 10^8}{2.5 \times 1.6 \times 10^{-19}} = 497 \text{ nm}$$

The range of this basic emulsion could be extended by the addition of a dye, which requires an energy of less than 2.5 eV to liberate one of its electrons. In this case photons could liberate electrons from the dye, which then could be taken up by the silver atoms to produce the latent image. This technique is used to sensitise silver halide films to green, yellow, orange, red and infrared light, but in the latter case, thermal effects may cause excitation of the dye electrons and excessive fogging may result if careful temperature control is not maintained. The sensitisation of silver bromide to photons of longer wavelengths by a dye is illustrated in figure 5.8.

TYPES OF EMULSION

An emulsion which is sensitised by the use of dyes to the whole of the visible spectrum is known as a *panchromatic* emulsion.

68

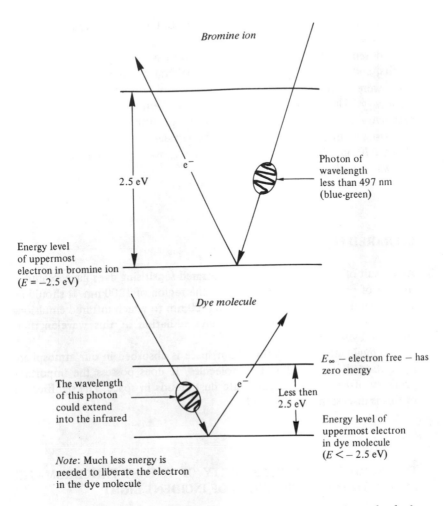

Figure 5.8 *Illustrating how electrons may be made available from dye molecules by photons of 'low' energy*

However, the first use of a sensitising dye was in 1882 when eosin was used to extend the sensitivity of an emulsion from the blue region of the spectrum into the green. Plates sensitised with eosin were described as 'isochromatic', even though the emulsion did not give an equal response to all colours as the name implies.

In 1884, erythrosin was used as a sensitising dye and improved the relation between the rendering of blue and green. Such plates were termed 'orthochromatic', meaning 'correct colour', but again this was an exaggeration.

Red sensitising dyes were also employed at this time but gave poor results, and it was not until 1936 that the first commercial panchromatic films were marketed. Numerous dyes are now employed to increase red sensitivity. The main differences between them lie in the ratio of red sensitivity to overall sensitivity, and in the position of the long-wave cutoff. Usually in panchromatic films the red sensitivity is extended up to about 670 nm since the sensitivity of the human eye is very low beyond this wavelength.

INFRARED PHOTOGRAPHY

As a result of successive discoveries, infrared sensitising dyes have extended the use of photographic emulsions to the region of 1200 nm. It should be noted that there is a practical limit of 1400 nm to which infrared emulsions could theoretically be sensitised, since radiation at this wavelength is absorbed by water.

Although infrared radiation from space is absorbed in our atmosphere by carbon dioxide and water molecules, it does possess the important property of being able to penetrate dust clouds in space. The significance of this is discussed in chapter 11.

THE VARIATION OF SENSITIVITY OF A FAST PANCHROMATIC FILM WITH THE WAVELENGTH OF INCIDENT LIGHT

Although a panchromatic film is sensitised to the whole of the visible spectrum, the response across the waveband is not uniform. The response is still most marked in the blue region, and then falls off as the wavelength increases. This is shown in figure 5.9. Included in the same figure is the spectral sensitivity curve for the human eye, and if it is desired that the photograph should respond to colours in exactly the same manner as the eye, then this is achieved by using *colour filters* in front of the camera lens. The filters restrict the intensity of those spectral regions which are detected more strongly by the film than the retina. It should be noted that we are still discussing black and white photography and that colour filters are a means of controlling the reproduction of colours in terms of greys.

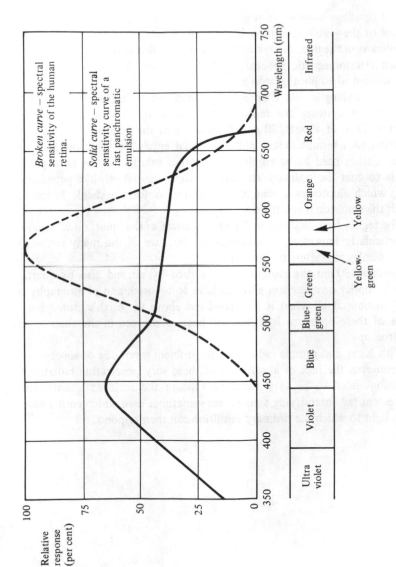

Figure 5.9 *Showing the spectral sensitivity curve of a fast panchromatic emulsion and comparing it with the spectral sensitivity of the human eye*

Broken curve – spectral sensitivity of the human retina.

Solid curve – spectral sensitivity curve of a fast panchromatic emulsion

THE RESPONSE OF SILVER BROMIDE EMULSION TO WAVE-LENGTHS SHORTER THAN THE VISIBLE REGION

The basic silver bromide film is sensitive not only to the blue and violet regions of the visible spectrum, but also to shorter wavelengths, including the ultra violet region, X-rays and gamma rays, down to the limits of the known electromagnetic spectrum. However, there are special problems encountered when photographs are taken at such wavelengths.

With wavelengths less than 330 nm where glass absorbs the radiation, glass optical systems are replaced by quartz or fluorite transmission systems. Then at about 230 nm the gelatine of the emulsion absorbs the radiation to a serious extent and so special emulsions with low gelatine concentration need to be employed. Another solution to the same problem is to coat the ordinary emulsion with a mineral oil, like petroleum jelly, which fluoresces during the exposure to form a visible image to which the emulsion is sensitive.

The region from 360 nm to 230 nm, known as the 'quartz u.v.' region, is particularly important in spectroscopy because of the many elements which display characteristic lines in it.

Beyond 180 nm the radiation is absorbed by air and also by quartz. Fluorite optics or reflection gratings have to be used, and photography is only possible at all when it is carried out above the earth's atmosphere. Some of these problems have already been mentioned in the chapter on spectroscopy.

With X-ray and gamma radiation, the problem is not one of absorption, but concerns the lack of absorption of these very penetrating radiations. Emulsions need to be very thick to capture the incident quanta, and fluorescent-salt intensifying screens are sometimes used which emit a blue-green light to which the ordinary emulsion can then respond.

6 Photoelectric Detection of Electromagnetic Radiation

A photoelectric transducer is a device which converts an incoming light signal into a measurable electrical response. There are a number of photoelectric devices which are used in astronomy, including photoemissive cells, photomultipliers and image intensifiers. Photoelectric detectors provide the most accurate means of measuring the intensity of stellar radiation and also the most accurate means of analysing stellar spectra.

A photoemissive cell is shown in figure 6.1. The cathode consists of a metal plate which has a surface coating of a photosensitive material, and is maintained at a negative potential. When light of a frequency which is greater than the threshold frequency of the photosensitive material is incident on the cathode, electrons are emitted from its surface, and are attracted to the anode. A current of the order of micro-amps may be recorded in the external circuit, and this current is found to vary in direct proportion to the intensity of incident radiation. The cathode is usually concave or 'V'-shaped, to form a simple light trap, so that if a photon is not absorbed on its initial impact with the cathode, it may be reflected on to another part of the surface and subsequently absorbed.

THE PHOTOEMISSIVE SURFACE

The mechanism of the photoelectric effect has already been mentioned. For a photon to liberate an electron from the photoemissive surface, the photon must possess an energy which is at least equal to the work function of the surface. If the work function of the photoemissive surface is designated as W Joules or ϕ electron volts then the minimum energy which a photon must posesses to be detected is given by

$$hf_0 = W = e\phi$$

where f_0 is the 'threshold' frequency for photoelectric emission. The 'threshold' wavelength for photoelectric emission is given by

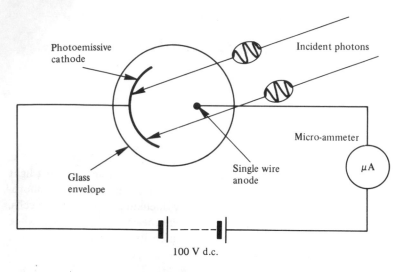

Figure 6.1 *Generalised diagram of a photoemissive cell*

$$\lambda_0 = \frac{c}{f_0} = \frac{hc}{e\phi}$$

If a photocell is to be sensitive to visible light, then the photosensitive surface of the cathode must have a threshold wavelength longer than 400 nanometres, and this requires that the work function of the surface material is less than 3.1 electron volts. That is

$$\phi = \frac{hc}{e\lambda} = \frac{6.62 \times 10^{-34} \times 3 \times 10^8}{1.6 \times 10^{-19} \times 400 \times 10^{-9}} = 3.1 \text{ eV}$$

Only the alkali metals possess work functions which are less than this. If a photocell is required to respond to the whole of the visible spectrum then the threshold wavelength increases to about 750 nanometres and the corresponding work function falls to 1.65 eV. No pure metals have such a low work function, but composite substances have been manufactured which do, and these can even be sensitised to detect infrared radiation. *Composite cathodes* are formed by evaporating one metal on to another, and it has been found that a monatomic layer of one metal lying on another, acting as a base, can have a work function much lower than that possessed by either metal.

74

SPECTRAL SENSITIVITY OF A PHOTOEMISSIVE SURFACE

The 'sensitivity' of a photocathode surface is defined as the ratio of the current emitted to the power carried by the radiation falling upon it; that is

$$S = \frac{I}{P}$$

The units of sensitivity are amp Watt^{-1}. Figure 6.2 shows the spectral response curves of three types of composite cathode in which sensitivity is plotted against the wavelength of incident light.

The antimony-caesium photocathode is highly sensitive, but its sensitivity does not extend into the red part of the visible spectrum. The bismuth-oxygen-silver-caesium composite covers the complete visual spectrum and the casesium–oxygen–silver composite extends the use of photocells into the infrared region.

THE QUANTUM EFFICIENCY OF A PHOTOEMISSIVE SURFACE

Not all the photons which fall on to a photocell result in the release of photoelectrons, even when the frequency of radiation is greater than the threshold frequency. Some of the radiation is reflected or absorbed by the glass envelope of the photocell, and the absorption of photons by glass prohibits the use at all of photoemissive cells in the ultra violet region. Some photons are reflected by the cathode, and others pass through the photoemissive surface without interaction.

The ratio of the number of electrons emitted for the number of photons incident is called the *quantum efficiency*, E

$$E = \frac{n_e}{n_p}$$

where E is the quantum efficiency, n_e is the number of electrons emitted and n_p is the number of incident photons.

It is possible to calculate the quantum efficiency from the sensitivity as follows

Sensitivity is defined as $\dfrac{\text{photoelectron current}}{\text{incident radiative power}}$

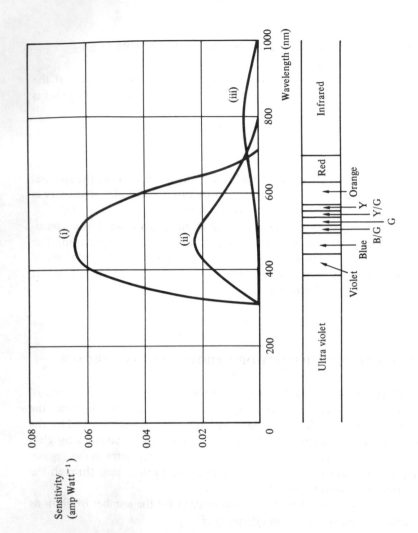

Figure 6.2 *Spectral response curves for three composite photocathodes*

(i) Antimony–caesium

(ii) Bismuth–oxygen–silver–caesium

(iii) Caesium–oxygen–silver

The photoelectron current is the amount of charge flowing in unit time, which is equal to the number of photoelectrons multiplied by the electronic charge, flowing in unit time

$$\text{Photoelectron current} = \frac{\text{charge}}{\text{time}} = \frac{n_e \times e}{t}$$

The incident radiative power of illumination is equal to the energy incident in unit time, and the energy is equal to the number of photons involved multiplied by the energy of each photon, which is hf

$$\text{Incident radiative power} = \frac{\text{energy}}{\text{time}} = \frac{n_p \times hf}{t}$$

Hence

$$\text{Sensitivity} = \frac{\text{photoelectron current}}{\text{incident radiative power}} = \frac{n_e}{n_p} \times \frac{e}{hf}$$

But n_e/n_p is the quantum efficiency, therefore

$$E = \left(\frac{hf}{e}\right) \times S$$

or in terms of wavelength

$$E = \left(\frac{hc}{e\lambda}\right) \times S$$

THE THEORETICAL MAXIMUM VALUE OF SENSITIVITY

The maximum efficiency that a photosensitive surface could theoretically achieve is one electron liberated for every incident photon, whereupon $E = 1$.

The maximum value of sensitivity which would be achieved under this hypothetical condition would depend upon the wavelength of the radiation. For blue-green light, for example, of wavelength 500 nm the maximum possible sensitivity would be calculated as follows

$$S = \frac{e\lambda}{hc} \times E$$

$$= \frac{1.6 \times 10^{-19} \times 500 \times 10^{-9}}{6.62 \times 10^{-34} \times 3 \times 10^8} \times 1 \approx 0.4 \text{ amp Watt}^{-1}$$

It is possible to calculate the theoretical maximum values of sensitivity for all wavelengths, and the last equation shows that the theoretical maximum value is directly proportional to wavelength. A plot of the theoretical maximum values of sensitivity versus wavelength is shown in figure 6.3.

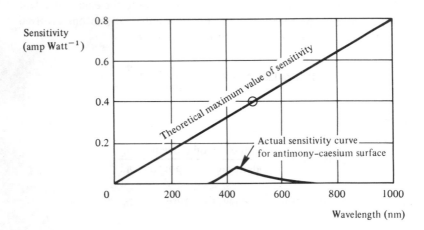

Figure 6.3 *A plot of the theoretical maximum value of sensitivity versus wavelength for a photoemissive surface (on the same axes is plotted the actual sensitivity curve for an antimony-caesium surface)*

COMPARISON OF QUANTUM EFFICIENCIES OF VARIOUS DETECTORS

It is possible to calculate the quantum efficiency of a photoemissive surface using the data provided in figure 6.3. For the antimony-caesium surface, the peak sensitivity of 0.065 amp Watt^{-1} is reached at a wavelength of 460 nanometres. Using

$$\text{Quantum efficiency} = \frac{hc}{e\lambda} \times S$$

we have

$$\text{Quantum efficiency} = \frac{6.62 \times 10^{-34} \times 3 \times 10^8}{1.6 \times 10^{-19} \times 460 \times 10^{-9}} \times 0.065 = 0.175$$

This means that for this particular surface and this particular wavelength, the quantum efficiency is 0.175 or 17.5 per cent. Generally, over their working range of wavelengths, photoemissive surfaces will have a quantum efficiency of between 10 and 20 per cent.

This figure is higher than that of 10 per cent obtained for the eye, and much higher than the figure of 0.16 per cent obtained for a 'fast', photographic emulsion. We saw in the case of the photographic emulsion that the low quantum efficiency is compensated for by the amplification which occurs during the development process. Currents from photoemissive surfaces may also be amplified if the photoemissive surfaces are incorporated into photomultiplier tubes. Photomultiplier tubes are discussed next.

THE NEED FOR AMPLIFICATION

The intensity of starlight reaching earth is very weak. A rough figure for a star which appears to the unaided eye to be easily visible is 10^{-5} Watts per square metre. If one now considers that the area of the cathode in a photocell is about 1 square centimetre, then the energy that a cathode receives from such a star is about 10^{-9} Watts. Using again the data for an antimony–caesium surface given in figure 6.3 we find that at the wavelength giving maximum sensitivity, a current of 0.065 amp is obtained per Watt. The 10^{-9} Watts from the star will therefore yield a photoelectric current of 6.5×10^{-11} amp. Such a current is too feeble to be useful and a photocell is only of value in astronomy if some means for current amplification is provided.

'DARK' CURRENT

Photocells give rise to weak currents, up to 10^{-9} amp, even when their cathodes are not illuminated. This 'dark' current is due to the 'thermionic' emission of electrons from the cathode. Thermionic emission is a phenomenon which occurs when a metal surface is heated, and the kinetic energy of the electrons contained within the surface increases, to an extent where some of the electrons escape. The rate at which electrons escape from the surface increases with temperature.

Since the energy required to free an electron from the photoemissive surface is very low, thermionic emission occurs even at room temperature, and the 'dark' current for an antimony–caesium surface of 1 square centimetre is about 3×10^{-10} amp at a temperature of 300 K.

In order to reduce the 'dark' current to a minimum, photoemissive surfaces are cooled. An antimony-caesium surface is usually operated at below 273 K, and the normal operating temperature for a caesium-oxygen-silver surface is 190 K.

THE PHOTOMULTIPLIER

The purpose of a photomultiplier is to 'multiply' the number of photo-electrons that have been emitted by a photoemissive surface, so that the photomultiplier cell produces an electric current which is easily measure-able. Figure 6.4 shows schematic representations of two types of photo-multiplier tube. The only significant difference between the two types is that the first has 'grid' dynodes, whereas the second has 'solid' dynodes.

Both types of photomultiplier consist of high-vacuum tubes in which photons enter through a glass window at the top and strike a very thin layer of photosensitive material, usually antimony-caesium, which is the photocathode. About one photoelectron is released from the photocathode for approximately every ten photons which fall on to it. Since the photo-cathode is very thin the photoelectrons emerge on the underside, and are attracted by the positive potential on the first dynode. Photomultipliers usually have between 9 and 14 dynode stages, each at a potential which is 100 volts higher than the previous stage.

In the grid type of tube, each dynode stage comprises a photoemissive surface on a metal grid, having a secondary emission coefficient of, say, 3, 4 or 5, which means that 3, 4 or 5 secondary electrons are released when an initial electron strikes. Thus a photoelectron emitted by the cathode is accelerated by the electric field to the first dynode where it produces a bunch of secondary electrons. These electrons pass through the grid and are accelerated to the next dynode, where they in turn produce more electrons, and this process repeats through each dynode stage.

In the solid dynode type of tube, the electrons do not pass through the dynodes, but are focused from one dynode surface on to the next by electrostatic fields.

After the electron pulse has traversed the complete dynode structure and finally reached the anode, the gain in the number of electrons may be as high as 10^9. In this way the incident photon gives rise to a burst of electrons at the anode where an electrical pulse is produced for further analysis.

The time needed to carry the pulse through the dynode system is called the 'rise time' and is generally about 10^{-8} or 10^{-9} second. This very fast

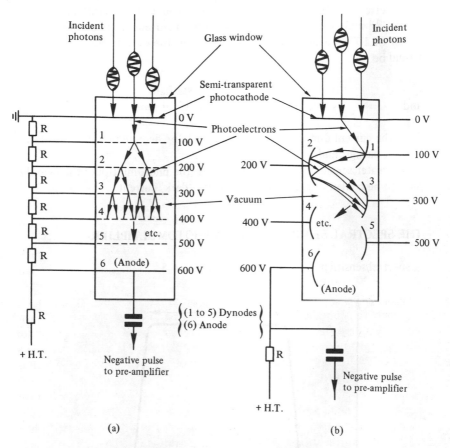

Figure 6.4 *Schematic representations of two types of photomultiplier tube. Type (a) has grid dynodes and (b) has solid dynodes*

response characteristic contributes greatly to the effectiveness of the photomultiplier as an instrument for measuring the intensity of starlight.

The *amplification* produced by a photomultiplier is equal to the 'secondary emission coefficient' raised to the power of the number of dynode stages. For example, if a photomultiplier has nine dynode stages and each stage has a secondary emission coefficient of 5, then the amplification is given by 5^9, which is approximately one million. If we consider the photocathode current of 6.5×10^{-11} amp which is obtainable from a bright star, then this could be amplified to 65 microamps by a nine-dynode stage photomultiplier.

We have seen on page 79 that a current of 6.5×10^{-11} amp is obtained for 'weak' starlight from an antimony–caesium cathode of area 1 square centimetre and that this same surface would give a 'dark' current of 3×10^{-10} amp at room temperature. If both currents were amplified by a factor of 10^6 in a photomultiplier tube then the output values of current would be

65 μA for 'weak' starlight

and   300 μA for the 'dark' current

Thus the current due to starlight could easily be detected against the 'dark' current background.

If greater sensitivity was required then the photomultiplier tube would have to be cooled to reduce the size of the 'dark' current.

THE SPECTRAL SENSITIVITY OF A PHOTOMULTIPLIER

A spectral sensitivity curve is shown in figure 6.5.

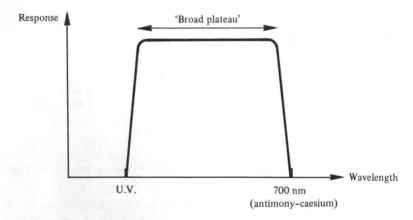

Figure 6.5 *A spectral sensitivity curve for a photomultiplier*

As regards the spectral sensitivity of the photomultiplier tube, the initial quantum efficiency of the photocathode surface is of little importance, as a result of the large amplification which subsequently takes place. The spectral response curve therefore has a very broad plateau, which is cut off at high frequency at the limit imposed by the glass envelope absorbing ultra violet light, and at low frequency by the 'minimum quantum energy' threshold required to liberate electrons from the photocathode surface, which occurs at a wavelength of 700 nm for an antimony-caesium cathode.

This broad spectral sensitivity curve makes the photomultiplier tube the most important instrument for measuring stellar brightnesses, and for use with a spectrometer in the analysis of stellar spectra, since it gives an equal response to all wavelengths.

An arrangement for attaching a photomultiplier tube to a telescope is shown in figure 6.6. Such an arrangement is suitable for stars which are bright enough to be seen through the telescope easily. The telescope can be directed at the star by an observer looking through the telescope eyepiece at reflected light from the prism. At the same time, light transmitted through the prism can be focused on to the photocathode of the photomultiplier tube.

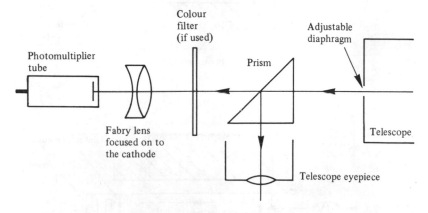

Figure 6.6 *Schematic arrangement for attaching a photomultiplier to a telescope*

THE IMAGE INTENSIFIER

In the photoelectric cell the electrons are accelerated and the current from the photocathode is measured. Any image on the photocathode surface is thereby lost. In an image intensifier however, the electrons released from the photocathode are accelerated, and detected when they fall on to a phosphor screen or photographic plate. The electrons are focused either magnetically or electrostatically so that the same image that was produced on the photocathode surface is now formed on the phosphor screen or photographic plate. Image intensifiers are shown in figure 6.7.

Image intensifiers are used in astronomical photography to provide enhanced sensitivity or to cut down a long exposure time which may otherwise be necessary. Image intensifiers incorporating phosphor screens

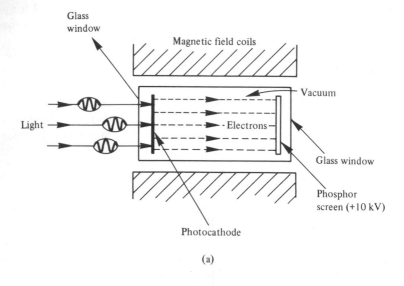

(a)

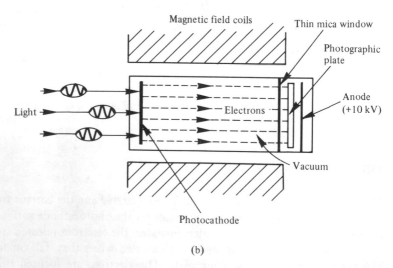

(b)

Figure 6.7 *Image intensifiers: (a) with a phosphor screen, (b) with a photographic plate*

have been found particularly valuable in measuring rapid variations in light intensity from certain celestial objects, such as 'pulsars' for example, which would prove impossible to detect from photographs.

It is possible to use image tubes in series to produce even greater amplification. An 'image tube scanner' at the Lick Observatory uses three image tubes in line to enhance the light in the spectra of faint stars. The final images are measured photoelectrically and recorded on magnetic tape for computer analysis. An arrangement for using image tubes in series is shown in figure 6.8.

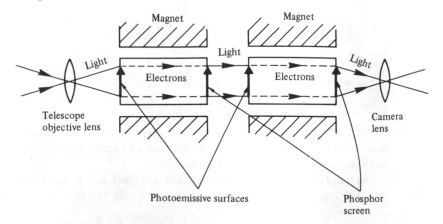

Figure 6.8   *A two-stage image tube (the final image is twice enhanced)*

PHOTOCONDUCTIVE CELLS

There are some substances which have the property that their electrical conductivity increases when they are illuminated with infrared light. Lead sulphide, indium antinomide and thallium sulphide are such substances, and are used to manufacture 'photoconductive cells'. Photoconductive cells give their greatest response to infrared radiation, and the 'illumination' falling upon such a cell can be measured by recording the current which it allows to pass when illuminated, and then extrapolating from a 'characteristic' curve of illumination versus current, for the particular cell in question.

The quantum efficiency of a photoconductive cell may be as high as 50 per cent and the infrared sensitivity may be up to 1000 times greater than a good photoemissive surface.

Cooling these detectors to very low temperatures increases the sensitivity, reduces noise, and in the case of lead sulphide improves the long-wavelength threshold.

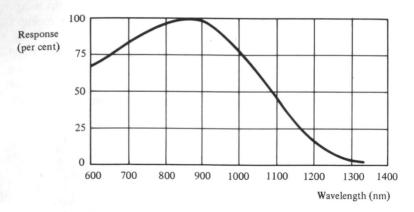

Figure 6.9 *The spectral response curve of a thallium sulphide cell*

Photoconductive cells are also manufactured from silicon semi-conductors, and silicon has the important advantage that its change of resistance is proportional to the light flux incident upon it.

The spectral response curve of a thallium sulphide cell is shown in figure 6.9. The high response to infrared radiation should be noted.

7 The Doppler Effect

The velocity of electromagnetic radiation in a vacuum is a constant which depends only upon the magnetic permeability and electrostatic permittivity of free space. This velocity is independent of any movement of the source or observer, but the same is not true of the wavelength and frequency of radiation. If the source is moving towards the observer there is an apparent increase in frequency and decrease in wavelength, and the converse is also true.

Consider figure 7.1 which shows a source of electromagnetic radiation S, which has a velocity v from left to right. Suppose the source is emitting a light wave of wavelength λ, and begins to radiate the wave at position x. Since this electromagnetic wave travels with the constant velocity c, the time taken for the front of the wave to reach point y is given by

$$t = \frac{\lambda}{c} \qquad (7.1)$$

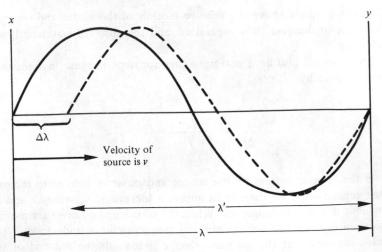

Figure 7.1 *Illustrating the Doppler effect*

However, in this short time-interval, the source will have moved a small distance $\Delta\lambda$ such that

$$\Delta\lambda = vt = \frac{v\lambda}{c} \qquad (7.2)$$

and the latter part of the wave will be generated from a point which is closer to y than the point from which the front part of the wave was generated. The result is an apparent shortening of the wavelength to λ', such that

$$\Delta\lambda = \lambda - \lambda' \qquad (7.3)$$

Equation (7.2) is known as the Doppler formula and enables us to calculate the change in wavelength due to the relative movement of source and observer. Note that the equation is the same if the source is stationary and the observer is receding. Also, combining equations (7.2) and (7.3) we have

$$\frac{\Delta\lambda}{\lambda} = \frac{\lambda - \lambda'}{\lambda} = \frac{v}{c} \qquad (7.4)$$

In the example above the relative motion of the source and observer was so as to decrease their separation, and the observed wavelength was shortened.

There would also be a corresponding apparent increase in frequency which is given by

$$\frac{f}{\Delta f} = \frac{f}{f - f'} = \frac{v}{c} \qquad (7.5)$$

If the relative motion of the source and observer is so as to increase their separation, then there is an apparent increase in wavelength and an apparent decrease in frequency. When the source and observer are moving apart we say that the wavelength and frequency have undergone a 'red shift', meaning that the apparent change is towards the red end of the spectrum. Conversely, when the source and observer are moving together we say that there is a 'blue shift'.

THE RADIAL SPEED OF A STAR

The radial speed of a star is the speed with which it is approaching or receding from the earth, along the line of sight. It represents a resolved component of the stars actual or 'peculiar' velocity, which may be in any other direction. If a spectrogram of the moving star is taken in a way which allows a comparison spectrum of a laboratory source to be photographed immediately beneath it, then the relative positions of the spectral lines can be compared, and any Doppler shift can be measured. The comparison spectrum is an emission line spectrum of an element for which the wavelengths of the emission lines are well known.

Appearing in the stellar spectrum will be certain absorption lines characteristic of elements which are present in the stellar atmosphere, and those lines which correspond to the element which was chosen for comparison need to be identified. Light from the two sources is directed through the upper and lower halves of the same grating, and corresponding lines would normally appear one directly above the other, were it not for the Doppler shift in the stellar spectrum. The apparent shift in wavelengths $\Delta\lambda$ can be determined by comparison of the two spectrograms, and the radial speed of the star can then be calculated using the formula (7.4).

THE RELATION BETWEEN THE APPARENT SHIFT IN WAVELENGTH AND THE ORIGINAL WAVELENGTH

If we examine again equation (7.4), which is

$$\frac{\Delta\lambda}{\lambda} = \frac{v}{c}$$

we see that if the original wavelength has a large value, then the corresponding, apparent change in wavelength will also be large, since for any given source and observer v is constant. On a spectrogram, therefore, we find that for a given combination of observer and source, moving either together or apart, the Doppler shift in wavelength of a particular spectral line is large if the wavelength itself is large. Conversely, a spectral line of shorter wavelength will undergo a smaller Doppler shift.

If we consider two spectral lines which were emitted at wavelengths of, say, λ_1 and λ_2, and the Doppler shift of the first line is $\Delta\lambda_1$, then we have from equation (7.4) that the Doppler shift of the second line is $\Delta\lambda_2$, where

$$\Delta\lambda_2 = \frac{\Delta\lambda_1}{\lambda_1} \times \lambda_2$$

RED SHIFT AS A DISTANCE INDICATOR

In 1912 V. M. Slipher had been measuring the radial velocities of galaxies by observing the Doppler shift of their spectral lines and he found that all distant galaxies exhibit a red shift, which indicates that they are receding from us. In 1929 E. Hubble found that there exists a linear relationship between the velocity of recession of a galaxy as measured by the red shift, and its distance from our own galaxy. Thus the further away a galaxy is, the faster it is receding. This is known as *Hubble's law*.

Most recent evidence supports Hubble's law although allowances must be made for the problems involved in determining the distances to the galaxies. Hubble based his assessments of distance on the assumption that the faintness of galaxies indicates their distance from us. This is a reasonable assumption but does not provide for very accurate measurement. The techniques of distance measurement employed in astronomy are discussed in the next chapters. Figure 7.2 gives examples of the velocity of recession of some galaxies, together with the distances of these galaxies from the earth.

Galaxy in constellation	Distance (light years)	Velocity of recession given by red shift (km s^{-1})
Virgo	43×10^6	1200
Ursa major	560×10^6	15 000
Corona borealis	728×10^6	21 600
Bootes	1290×10^6	39 300
Hydra	1960×10^6	61 000

Figure 7.2 *The recession velocities of five galaxies*

For a given galaxy we can define a quantity known as the 'red shift' (z), such that

$$z = \frac{\Delta\lambda}{\lambda} \tag{7.6}$$

When a red shift occurs the apparent wavelength λ' will be greater than the 'true' wavelength λ, and we have

$$z = \frac{\Delta\lambda}{\lambda} = \frac{\lambda' - \lambda}{\lambda}$$

From equation (7.4) we have

$$z = \frac{\Delta\lambda}{\lambda} = \frac{v}{c}$$

so that the velocity of recession of a galaxy is given by

$$v = zc \tag{7.7}$$

We can express Hubble's law in terms of the 'red shift' such that

$$v = zc = Hr$$

where r is the distance to the galaxy, and H is 'Hubble's constant'.

Hubble's constant has been calculated to lie between the values 50 and 125 km s^{-1} Mpc^{-1}. The large uncertainty in its value is due to the uncertainties in measuring the distances to the galaxies. The quantity Mpc or megaparsec which appears in the units of Hubble's constant is a unit of distance and is discussed in the next chapter (1 Mpc = 3.086×10^{22} m). It is often convenient to use a value of 100 km s^{-1} Mpc^{-1} for the Hubble constant.

The reciprocal of Hubble's constant (H^{-1}) has the units of time. Using dimensional analysis on the equation

$$r = \frac{v}{H} = vH^{-1} \tag{7.8}$$

We see that the dimensions involved in the equation are

$$[L] = [LT^{-1}] \times [T]$$

where the last term gives the dimensions of H^{-1}.

THE SIZE AND AGE OF THE UNIVERSE

What exactly constitutes the universe, and questions relating to its age and size are problems of philosophy as well as science, but if we define the universe as being all that matter to which we theoretically could have access, if we travelled for a long enough time, then we can put a limit on its size.

We begin by examining the theory which states that the universe is expanding, which is a conclusion that can be reached from Hubble's law. It seems that all distant objects in the universe are receding from us, and since we have no reason to suppose that we are situated at the centre of the universe we also suppose that we are receding from them. This therefore leads us to assume that the whole universe is expanding.

As objects become more distant from us their velocity of recession increases and at a certain distance the speed of such objects will approach the speed of light. However, it is a fundamental law of physics that no object can travel faster than the speed of light, and so once objects approach such a speed relative to earth, they are lost to us for ever, since we could never envisage a means of travelling fast enough to catch them. We can therefore consider the distance at which objects approach the speed of light as providing a theoretical limit to the size of the universe as measured from the earth. The actual distance involved is calculated from equation (7.8) as follows

$$r = \frac{v}{H}$$

If $v = c = 3 \times 10^8$ m s^{-1} and H is taken as 100 km s^{-1} Mpc^{-1}, then we must first convert the units km s^{-1} Mpc^{-1} to s^{-1} before we can perform the division. Thus

$$H = 100 \text{ km s}^{-1} \text{ Mpc}^{-1} = 10^5 \text{ m s}^{-1} \text{ Mpc}^{-1} = \frac{10^5}{10^6} \text{ m s}^{-1} \text{ pc}^{-1}$$

Then, since 1 pc = 3.085×10^{16} m, we have

$$H = \frac{10^5}{10^6 \times 3.085 \times 10^{16}} \text{ m s}^{-1} \text{ m}^{-1} \text{ or s}^{-1}$$

Now dividing v by H we have

$$r = \frac{3 \times 10^8 \times 10^6 \times 3.085 \times 10^{16}}{10^5} \approx 10^{26} \text{ m}$$

and we find that the universe pertaining to any individual observer has a radius of 10^{26} metres, measured from the observer. (*Note:* It is possible to reach these 'inaccessible' regions by taking an indirect route via the origin of the big bang, provided one travels faster than the expansion velocity of the universe.)

The time of approximately 3×10^{17} s represented by H^{-1} also happens to be the 'age' of the universe, if we consider the beginning of the universe to have occurred at some instant when all matter was concentrated at some central point before it began its expansion outwards. We can understand why H^{-1} represents the age of the universe by considering figure 7.3.

The figure represents the distance measured to a distant galaxy from the 'centre of the universe', as a function of 'absolute' time (that is, measured since the beginning of the universe).

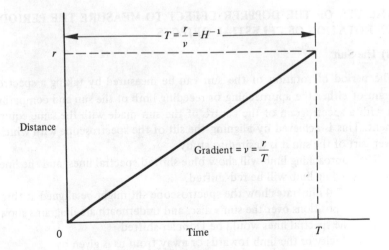

Figure 7.3 *Graph of distance from the centre of the universe to a distant galaxy versus absolute time*

If we assume that the expansion of the universe has been constant, then the 'average' velocity of expansion is given by the gradient of the graph. If the time T represents the present time, then the value of T is given by

$$T = \cfrac{r}{\text{gradient of line}} = \frac{r}{v}$$

but from Hubble's law $r/v = H^{-1}$, and therefore the time during which the galaxy has been involved in the expansion is H^{-1}, which is the age of the universe.

This value for the age of the universe shows reasonable agreement with other estimates based upon radioactive dating of meteorite samples, which are believed to be the oldest rocks currently available, and other techniques, based upon assumptions concerning the life cycles of stars. It should be noted that these estimates are subject to large inaccuracies, and therefore do not provide conclusive evidence that the red shift of distant galaxies does in fact indicate an expanding universe, although this is the most popular theory.

It should be noted that the calculation to find the age of the universe is over-simplified, since it takes no account of the gravitational attraction between the matter of the universe, which will slow down any expansion.

THE USE OF THE DOPPLER EFFECT TO MEASURE THE PERIODS OF ROTATION OF CELESTIAL OBJECTS

(i) The Sun

The period of rotation of the sun can be measured by taking a spectrogram of either the approaching or receding limb of the sun and comparing it with a spectrogram of the centre of the sun made with the same equipment. This is achieved by aligning the slit of the spectroscope with whichever part of the sun it is desired to study.

The approaching limb will show blue-shifted spectral lines, and the lines of the receding limb will be red-shifted.

Figure 7.4 illustrates how the spectroscope slit might be aligned at three successive positions over the sun's disc, and underneath are diagrams showing how the spectral lines would be Doppler-shifted.

The velocity of the limb towards or away from us is given by

$$v = c \; \frac{\Delta\lambda}{\lambda}$$

and if the spectroscope was aligned, say, at the outer edge of the sun's disc and the radius of the sun, r, is known then v represents the tangential velocity of the sun. The time period of rotation is given by

$$r = \frac{2\pi r}{v} = 2\pi r \times \frac{\lambda}{c\Delta\lambda}$$

The 'radius' of the sun is obtained from a knowledge of the angular diameter of the sun, and a knowledge of the distance between the earth and the sun. The angular diameter is found by taking bearings from a point at the earth's surface, to either edge of the sun. The distance between the earth and the sun is nowadays determined accurately by timing radar pulses, which are 'bounced' off the sun.

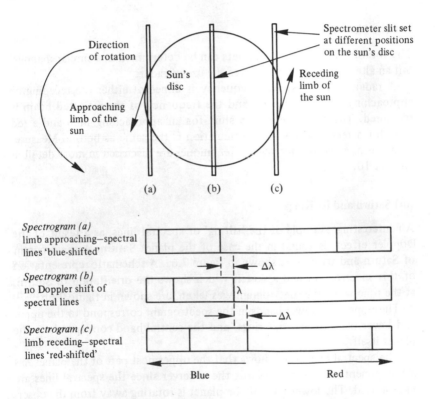

Figure 7.4 *Using the Doppler effect to measure the period of rotation of the sun*

The mean angular diameter of the sun is 39′ 59″ while the mean distance between the earth and the sun is 1.496×10^{11} metres. The radius of the sun can be calculated by simple trigonometry and is found to be 6.96×10^{8} m. These values are shown in figure 7.5.

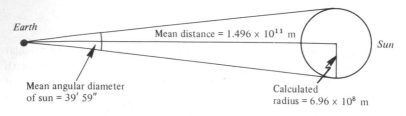

Figure 7.5 *Finding the radius of the sun*

(ii) The Planets

The rotation periods of the planets can be determined in a similar manner, but an alternative method involves the use of radar.

A radar beam of known frequency is aimed at either the receding or approaching limb of a planet, and the frequency of the reflected beam is measured. There will be a blue shift for an approaching limb and a red shift for a receding limb. The calculation is the same as before. Measurements using radar or 'microwave' techniques are discussed in more detail in chapter 10.

(iii) Saturn and its Rings

An interesting example of the 'tilting' of spectral lines, as a result of the Doppler effect, is found in the case of the planet Saturn. A spectrogram of Saturn and its rings is shown in figure 7.6a. A schematic representation of this spectrogram and a sketch which shows the orientation of Saturn, at the time when the spectrogram was taken, are shown in figure 7.6b.

The upper and lower bands in the spectrogram correspond to the upper and lower parts of Saturn's rings, and the central band corresponds to the planet itself.

The spectral line tilting shows that the uppermost part of the planet has a component of motion towards the observer since the spectral lines are blue-shifted. The lower part of the planet is rotating away from the observer and the spectral lines exhibit a red shift.

The shifts in the spectral lines representing the rings of Saturn are not so straightforward. The rings rotate in the same direction as the planet, as indicated by the blue shift of the spectral lines from the lower part of the rings. However, the inner edge of the rings gives a greater Doppler shift than the outer edge, which means that the inner edge is rotating faster than the outer edge. The rings do not therefore rotate as a solid body. Before information from spectrograms was available, James Clerk Maxwell

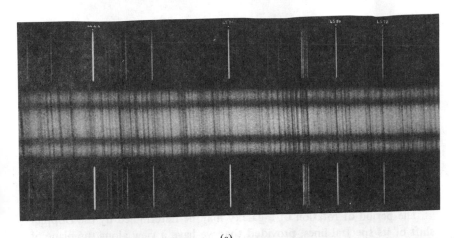

(a)

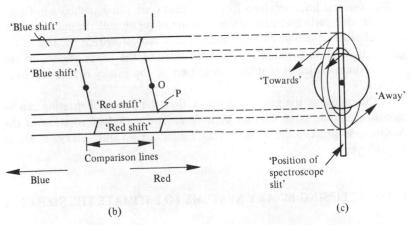

'Blue shift'

'Blue shift'

O

P

'Red shift'

'Red shift'

Comparison lines

Blue   Red

(b)

'Towards'

'Away'

'Position of
spectroscope
slit'

(c)

Figure 7.6 *(a) A spectrogram of Saturn and its rings (photograph reproduced by permission of The Royal Astronomical Society). (b) A schematic representation of the spectrogram of (a), and (c) showing the orientation of Saturn at the time of the spectrogram*

had already shown from theoretical considerations that owing to tidal forces the rings of Saturn must be composed of separate 'moonlets' with those on the inside moving fastest.

The period of rotation of Saturn may be determined by measuring say the red shift between points O and P to give the quantity $\Delta\lambda$. The true value of λ may be obtained from a comparison spectrum which is shown above and below the spectrum of Saturn in figure 7.6a. The value for r in the equation

$$\tau = 2\pi r \times \frac{\lambda}{c\Delta\lambda}$$

is the radius of the planet which is known, and hence the period of rotation can be found.

(iv) Galaxies

Galaxies take various forms, but two galaxies of the 'spiral' type are shown in figure 7.7. Figure 7.7a shows a spiral galaxy perpendicular to its plane of rotation, while figure 7.7b shows another spiral galaxy along its plane of rotation.

The period of rotation of a galaxy may be determined by the Doppler shift of its spectral lines, provided that we have a view along the plane of rotation. The slit of the spectrometer must be aligned along the major axis of the galaxy, as shown in figure 7.8a.

The spectral lines obtained from the galaxy are tilted as shown in figure 7.8b. In this particular illustration the upper regions of the spectral lines are inclined to the blue end of the spectrum, indicating that the top of the galaxy is rotating towards us. The red shift of the lower regions of the spectral lines indicates that the lower part of the galaxy is rotating away from us.

If the radius of the galaxy is known, then the period of rotation can be calculated in the same manner as before. The period of rotation of the Andromeda galaxy, which is a near neighbour to our galaxy, is about 3×10^8 years.

USING ECLIPSING BINARY SYSTEMS TO ESTIMATE THE SIZE OF A STAR

Accurate data on the dimensions of stars are difficult to obtain, but one useful method involves the study of eclipsing binary systems. A binary system or 'double star' consists of two stars in orbit about a common axis of rotation, with common orbital period, but usually having different orbital radii. Provided that we have an 'edge on view' of the orbits, we observe that periodically one star eclipses the other, and for the duration of the eclipse the spectral lines from the 'eclipsed' star disappear from a spectrogram. The time duration of such an eclipse would provide information about the diameters of the stars if it was known how fast the stars were moving in their orbits. This last piece of information may be obtained by measuring the Doppler shifts of the spectral lines.

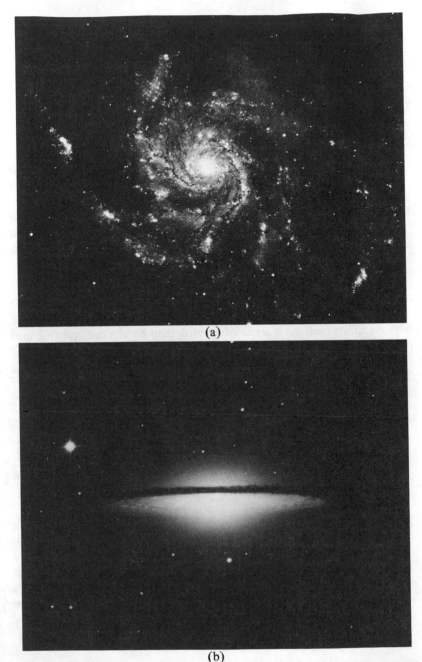

(a)

(b)

Figure 7.7 *(a) A spiral galaxy seen perpendicular to its plane of rotation (photograph from the Hale Observatories) – NGC 5457 – Galaxy in Ursa Major. (b) A spiral galaxy seen along its plane of rotation (photograph from the Hale Observatories) – M 104 – Galaxy in Virgo*

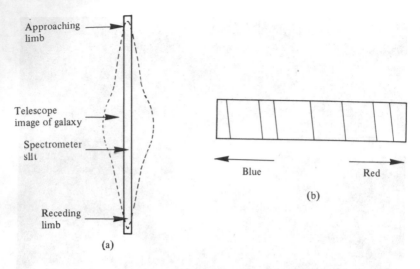

Approaching
limb

Telescope
image of galaxy

Spectrometer
slit

Receding
limb

(a)

Blue   Red

(b)

Figure 7.8   *(a) Position of spectrometer slit. (b) Typical spectrogram*

For a given spectral line which is emitted by both stars, the Doppler
shift manifests itself on a series of spectrograms by the line splitting into
two components, one generated by each star, and these two components
cross one another repeatedly as the stars go round in their orbits. This is
illustrated in figure 7.9.

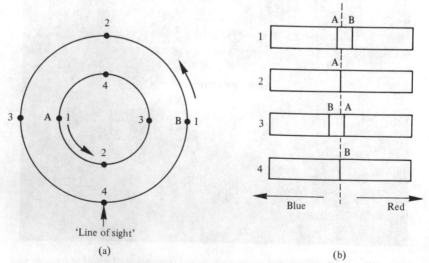

Figure 7.9   *(a) Showing the orbits and relative positions of stars A and B. (b) Illustrating the relative movements of the spectral lines*

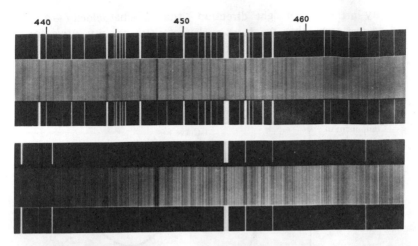

Figure 7.10 *Spectrogram of a binary system in Ursa Major, showing the doubling of lines (photograph by permission of The Royal Astronomical Society) — wavelength given in nanometres*

Figure 7.10 shows actual spectrograms for a binary system in Ursa Major. The lower spectrogram illustrates the doubling of the spectral lines when both stars are in view.

The star with the larger orbit travels faster, and the Doppler shift of its spectral line component will be greater than for the other star. The period of orbital rotation of either star is equal to the time taken by the oscillating spectral lines to complete one cycle. The maximum value of $\Delta\lambda$ for each star, corresponding to those positions in the orbits when the stars are moving directly towards or away from the earth, may be calculated from the spectrograms. The orbital speed of either star can then be calculated using

$$v = \frac{c\Delta\lambda}{\lambda}$$

where v is the orbital speed.

VELOCITY CURVES FOR SPECTROSCOPIC BINARIES

The orbital periods of spectroscopic binaries are typically from a few hours to a few months, since the two stars orbit their centre-of-mass closely. If over the time period for a complete orbit, values of the component of velocity in the 'line of sight' direction — that is, radial velocity — are plotted against time, then we have a characteristic velocity curve for the system. Such a curve is shown in figure 7.11.

The numbers 1 to 4 on this diagram correspond to the numbers on figure 7.9 which show the relative positions of the stars.

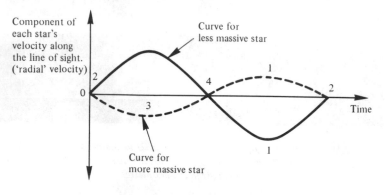

Figure 7.11   *A velocity curve for a spectroscopic binary system*

The curve with the larger amplitude is for the less massive star, since this star has the larger orbit and possesses the greater speed.

The curves are sinusoidal if the orbits are circular, since for an object moving in a circle the component of velocity in one particular direction is a sine function.

The maximum values of each curve correspond to the star moving directly towards or away from earth.

LIGHT CURVES FOR SPECTROSCOPIC BINARIES

The light output of a binary system can be monitored using a photomultiplier tube, attached to a telescope.

The time period of the orbits, and then the circumference of each orbit, could be found by plotting a 'light curve' for the binary system. The light output from the system is measured over a period of time, and a typical light curve is shown in figure 7.12.

The time period of the orbits can be read off the x axis after one complete cycle, and the circumferences are found using

$$\text{Circumference} = \text{Orbital speed} \times \text{Time period of orbit}$$

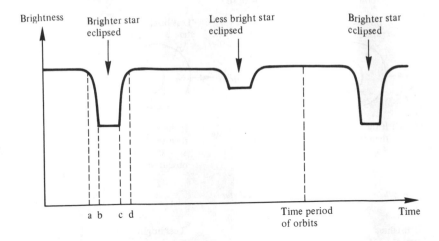

Figure 7.12 *Light curve for an eclipsing binary system − taken over one orbital period*

The stars in a binary system may differ in size and in brightness, and the light curve is characterised by two 'troughs' representing decreases in total light output for the system. The deeper trough occurs when the brighter of the two stars is eclipsed by the other, and this is true regardless of whether the brighter star is the larger or smaller of the two. The shallower trough occurs when the less bright star is eclipsed by the brighter star. This is illustrated in figure 7.13.

The diameters of the stars can be found from the time intervals ab and ac shown on figure 7.12. This particular 'trough' is shown again in figure 7.14, beneath a diagram showing the corresponding position of the two stars.

At time a the stars are beginning to eclipse, but the light intensity is still maximum. Between times a and b the extent of the eclipse increases and the total light output decreases gradually, as is shown on the graph. At time b the eclipse is total and the light output is a minimum. The eclipse remains total for the time interval bc. Between times c and d the stars are 'separating' and the total light output increases, until at time d the light intensity is a maximum again.

Let the velocities of the large and small star be V_L and V_S respectively. If we consider the large star to be fixed then the relative velocity of the small star as it passes across it, is given by

$$\text{Relative velocity} = V_L + V_S$$

(a) When the small star eclipses the large star

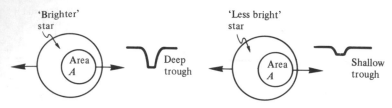

(i) If the large star is the brighter then the 'trough' is deep, as area A of the bright star is obscured

(ii) If the small star is the brighter then the 'trough' is shallow, as area A of the less bright star is obscured

(b) When the large star eclipses the small star

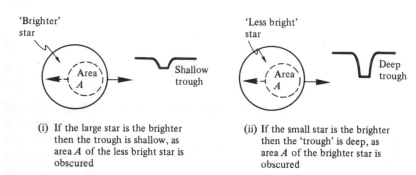

(i) If the large star is the brighter then the trough is shallow, as area A of the less bright star is obscured

(ii) If the small star is the brighter then the 'trough' is deep, as area A of the brighter star is obscured

Figure 7.13 *Illustrating how the depth of a 'trough' in a light curve is determined by whether it is the brighter or less bright star which is being eclipsed*

Considering the large star to be fixed, we have that in the time interval ab, the small star has moved a distance equal to its own diameter. The diameter of the small star is therefore given by

$$\text{Diameter of small star} = (V_L + V_S) \times \text{time interval ab}$$

Similarly, in the time interval ac, the small star has moved a distance equal to the diameter of the large star. The diameter of the large star is therefore given by

$$\text{Diameter of large star} = (V_L + V_S) \times \text{time interval ac}$$

Note: We can ignore any parallax effects due to the separation of the stars, since this distance is negligible compared with the distance of the stars from earth.

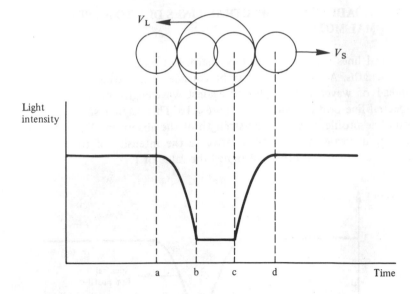

Figure 7.14 *Showing how the diameters of the stars are found from a trough in the light curve*

If we do not have an 'edge on' view of an eclipsing binary system, so that the eclipses we observe are partial and never total, then the light curve exhibits 'V'-shaped troughs. This is illustrated in figure 7.15.

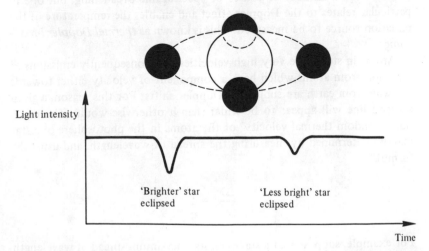

Figure 7.15 *A light curve for a binary system which exhibits partial eclipses*

THE BROADENING OF SPECTRAL LINES DUE TO RANDOM THERMAL MOTION

Spectral lines are never infinitely sharp, but exhibit a narrow spread of wavelengths. A spectral line may be considered as having a 'profile' if the spread of wavelength is plotted against wavelength intensity. A typical spectral line profile is shown in figure 7.16. This diagram shows an absorption line profile, in which the strength of the absorption line is measured by the decrease in intensity, relative to the intensity of the continuum radiation which is arbitrarily assigned the value of 1.

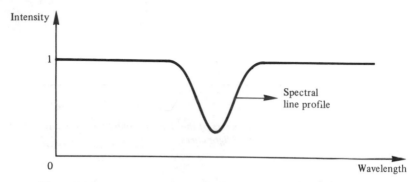

Figure 7.16   *The 'profile' of a spectral line in an absorption spectrum*

There are a number of causes of spectral line broadening, but one in particular relates to the Doppler effect and enables the temperature of the radiation source to be measured. This is known as *thermal Doppler broadening*.

Atoms in stars have very high velocities and consequently emissions of radiation from atoms which have a component of velocity either towards or away from earth are subject to Doppler shifts. For this reason, a given spectral line will appear to be wider than it otherwise would. The maximum 'random thermal velocity' of the atoms in the photosphere of a star may be determined by measuring the spread of wavelength, and using the formula

$$\frac{\Delta\lambda}{\lambda} = \frac{v}{c}$$

For example, suppose that a star exhibits a maximum spread of wavelength due to thermal Doppler broadening of 2.94×10^{-11} m, for the helium yellow line of wavelength 5.88×10^{-7} m. The maximum velocity of the helium atoms is found as follows.

The total broadening is due to atoms moving with velocity components towards and away from the earth. We must consider only those atoms with velocity components towards us, or those with velocity components away from us. In this case the broadening due to either of these groups of atoms is

$$\frac{2.94 \times 10^{-11}}{2} \quad \text{metres}$$

and the maximum thermal velocity of the helium atoms is

$$v = \frac{\Delta\lambda}{\lambda} \times c = \frac{2.94 \times 10^{-11} \times 3 \times 10^8}{2 \times 5.88 \times 10^{-7}} = 7.5 \times 10^3 \text{ m s}^{-1}$$

The thermal velocity of particles within a gas depends upon the temperature of the gas, and the temperature of a stellar atmosphere may be determined by measuring the spread of a spectral line due to thermal Doppler broadening.

From the kinetic theory of gases we have that the mean kinetic energy of a gas particle is given by the equation

$$\frac{\overline{mv^2}}{2} = \frac{mV_{rms}^2}{2} = \frac{3kT}{2}$$

where m is the mass of the particle, T is the absolute temperature, k is the Boltzmann constant $= 1.38 \times 10^{-23}$ Joule K^{-1}, and V_{rms} is the 'root mean square' value of velocity for all of the gas particles.

When we consider the mean kinetic energy of a gas particle it would be inappropriate to use the mean value of velocity, since the particles are moving in every direction and the mean value of velocity is therefore zero. However, if all components of velocity in one direction are designated as negative, then the squares of all the velocities are positive. It is then possible to find the mean value of the 'squared' velocities, and taking the square root of this quantity gives the appropriate value of velocity to fit into the above equation.

The value of the rms value of velocity is given by

$$V_{rms} = \frac{1}{\sqrt{2}} \times V_{max} = 0.707 \ V_{max}$$

Thus the procedure for determining the temperature of a stellar atmosphere is

(i) Analyse a spectral line to measure the amount of thermal Doppler broadening.

(ii) Divide the total amount of broadening by 2 to obtain the value of $\Delta\lambda$ to be substituted into the equation

$$v = c \ \frac{\Delta\lambda}{\lambda}$$

and in this way calculate v, which is the maximum velocity of a gas particle.

(iii) Find the root mean square value of velocity using

$$V_{rms} = 0.707 \ V_{max}$$

(iv) Finally, calculate the temperature of the gas using

$$T = \frac{m \ V_{rms}^2}{3k}$$

where m is the mass of the atom or molecule which emits the spectral line.

It should be noted that since

$$V_{rms} = \sqrt{\left(\frac{3kT}{m}\right)}$$

it follows that the root mean square velocities of heavier elements are less than the values for lighter elements. The consequence of this is that for a given temperature the thermal Doppler broadening for heavier elements is less than that for lighter elements.

Worked Example

As a worked example we can estimate the extent of thermal Doppler broadening on the Balmer α-line of neutral hydrogen, of wavelength 656.3 nm, when the temperature of the gas is 6000 K. Boltzmann's constant = 1.38×10^{-23} Joule K^{-1}, and the rest mass of a proton ≡ mass of a hydrogen atom = 1.67×10^{-27} kg.

Answer

The rms value of the velocity of the atoms is given by

$$V_{rms} = \sqrt{\left(\frac{3kT}{m}\right)} = \sqrt{\left(\frac{3 \times 1.38 \times 10^{-23} \times 6000}{1.67 \times 10^{-27}}\right)}$$

$$V_{rms} = 1.22 \times 10^4 \text{ m s}^{-1}$$

The maximum velocity of the atoms is

$$V_{max} = \frac{V_{rms}}{0.707} = 1.73 \times 10^4 \text{ m s}^{-1}$$

If we consider only those particles with velocity components towards the observer, then the broadening of wavelength is given by

$$\Delta\lambda = \frac{V_{max}\,\lambda}{c}$$

therefore

$$\Delta\lambda = \frac{1.73 \times 10^4 \times 656.3 \times 10^{-9}}{3 \times 10^8} = 0.038 \text{ nm}$$

This value of $\Delta\lambda$ gives the spread of wavelength from the centre of the profile to one edge. If we also consider particles with velocity components away from the observer, the total spread of wavelengths due to thermal Doppler broadening is given by

$$2 \times 0.038 = 0.076 \text{ nm}$$

8 The Stellar Magnitudes and Distances

DISTANCE DETERMINATION BY TRIGONOMETRIC PARALLAX

Trigonometric parallax provides a highly accurate method of determining the distance to the nearest stars. The technique is essentially the same as the terrestrial surveying method of triangulation, in which the distance towards a point is measured by taking sightings, and measuring the angles to the point from either end of a baseline of known length. The distance to the object is calculated by simple trigonometry.

As the earth orbits the sun, the nearer stars appear to move slightly with respect to the more distant stars behind them. This apparent, relative motion between near and distant stars is due entirely to the motion of the earth and is known as *parallax*. If the star is close enough for a 'parallatic shift' to be detected, then it is also possible to measure the apparent 'angular displacement' of the star relative to the background of 'fixed' distant stars, as the earth moves halfway around its orbit. In figure 8.1 the apparent angular displacement is the angle subtended by the earth as it moves from A to B in its orbit.

In order to measure θ a bearing to the star is recorded, and then after a time interval of 6 months the bearing is retaken. The difference between the bearings gives the angle. The baseline for triangulation in this case is 2 'astronomical units', where the astronomical unit (A.U.) is defined as the mean distance between the earth and the sun which is 1.496×10^{11} m. This distance is known to a high degree of accuracy and the distance d to the star can be calculated.

The angle p in figure 8.1 is known as the 'parallax' angle and astronomers often find it convenient to describe the distance to a star in terms of this angle. The nearest star to earth is Proxima Centauri which has a parallax angle of 0.76 arc seconds. (*Note:* One arc second is 1/3600th of a degree, and the arc second is a convenient unit for measuring these very small parallax angles. The symbol for arc seconds is ($''$) and 0.76 arc seconds, for example, is written as $0''.76$.)

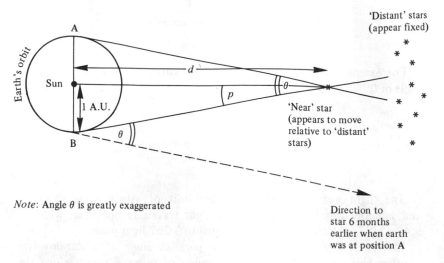

A

Earth's orbit

Sun

1 A.U.

d

p

θ

θ

B

'Distant' stars
(appear fixed)

'Near' star
(appears to move
relative to 'distant'
stars)

Note: Angle θ is greatly exaggerated

Direction to
star 6 months
earlier when earth
was at position A

Figure 8.1 *Trigonometric parallax*

THE PARSEC

Stellar distances are so large that it is convenient to use the unit of length known as the 'parsec' (pc), which is defined as the distance to a point in space, having a parallax angle of one arc second. This is illustrated in figure 8.2. One parsec is equal to 3.086×10^{16} m.

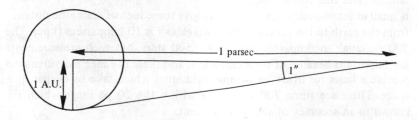

1 A.U.

1 parsec

1"

Figure 8.2 *The parsec*

Since a point with a parallax angle of $1''$ is at a distance of 1 pc it follows that a point with a parallax angle of $0''.5$ is at a distance of 2 pc, and so on.

It is therefore possible to convert a parallax angle, given in arc seconds, to a distance in parsecs using the relation

$$d \text{ (parsecs)} = \frac{1}{p \text{ (arc seconds)}}$$

For example, the distance to Proxima Centauri, which has a parallax angle of 0″.76, is given by

$$d = \frac{1}{0''.76} = 1.3 \text{ parsecs}$$

The 'light year' is a more familiar unit of astronomical distance, and is defined as the distance which light travels in one year, which is 9.46×10^{15} m. Hence one parsec is equal to 3.261 light years.

The procedure for measuring the parallax angle of a star involves photographing a star field from a number of selected points along the earth's orbit. A number of 'distant' stars with no detectable parallax are chosen as 'fixed' background markers and, during the course of one earth orbit, the star under observation appears to execute a cyclic motion. Corrections are made for any movement of the star relative to the background stars, which is not due to parallax such as the stars own 'peculiar' movement, and the parallax angle is calculated from measurements taken from the photographic plates.

Stellar parallaxes may be determined with a probable error of ± 0″.005 which means that parallax angles measured down to about 0″.01 are reliable, and this corresponds to a distance of about 100 pc. This distance is small in astronomical terms, since to give some idea of scale, the distance from the earth to the centre of our own galaxy is 10 kiloparsecs (kpc). The 100 pc range encompasses only some 6500 stars, but such measurements are important because of their reliability, and these distance measurements provide a basis for other measuring techniques which take us deeper into space. There are some 700 parallaxes within the 20 pc range which are known to an accuracy of about 10 per cent.

A SUMMARY OF DISTANCE UNITS

1 Astronomical Unit (A.U.) = mean earth–sun distance = 1.496×10^{11} m
1 Parsec (pc) = 3.086×10^{16} m = 2.063×10^5 A.U. = 3.261 Light Years
1 Light Year = 9.46×10^{15} m = 6.324×10^4 A.U. = 0.3066 pc.

DISTANCE DETERMINATION BY STATISTICAL PARALLAX

If it was possible to have a baseline for triangulation which was greater than the diameter of the earth's orbit, then parallax angles could be measured for stars which are at greater distances than 100 pc. A longer baseline is available if we take account of the motion of the sun through the nearby stars (that is, the 'peculiar' motion of the sun).

The direction of the sun's peculiar motion is determined by locating the direction in space in which the stars lying in that direction show, on 'average', velocities of approach to the sun. In the opposite direction are found stars which, on 'average', show velocities of recession. Since the average approach velocity of the stars 'ahead' of us is equal to the average recession velocity of the stars 'behind' us, we deduce that the average approach or recession velocity gives the peculiar motion of the sun among the other stars. This work involves detailed statistical studies, but it is established that the sun is travelling towards the constellation Hercules with a velocity of 20 km s^{-1} or 4.1 A.U. yr^{-1}. We therefore have a 'baseline' which increases by 4.1 A.U. in every year, and which over a 10 year period should allow distance measurements up to 2000 pc. Unfortunately it is not possible to use this long baseline to measure to 'individual' stars at such large distances, because each star has its own peculiar motion which is indeterminable at distances beyond 100 pc.

(*Note:* The peculiar motion of a star can be determined after the earth has made one complete solar orbit, when any shift of the star, in that 1 year period, relative to the background stars, can be analysed.)

However, the 'increasing' baseline is of use in determining the distances to 'clusters' containing a large number of stars, in which case it is assumed that the individual peculiar motions average to zero. For such a cluster it is then possible to deduce the mean or 'secular' parallax. This method of distance determination is known as 'statistical' parallax, and it can only be applied to a cluster of stars in which it is assumed that the distances between the stars in the cluster are very small compared with the distance of the cluster to the earth. Stellar clusters are discussed on page 131.

The method of statistical parallax, which employs the 'increasing' baseline, is illustrated in figure 8.3.

Statistical parallax is vital in establishing the distances to certain 'key' stars which exhibit special characteristics, such as fluctuations in brightness occurring with a precise period. These key stars fit into various classifications such as 'spectral class − A' stars, 'Cepheid variables' and 'R.R. Lyrae' type stars. Stars within each classification are assumed to be identical in all respects. Such stars can be recognised at very great distances from the earth, and far beyond the range of statistical parallax measure-

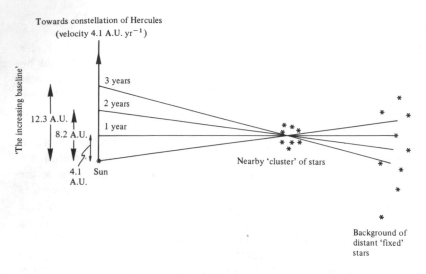

Figure 8.3   *Illustrating the method of statistical parallax*

ments. However, when these key types of star are found in clusters which are close enough for their distance to be determined by statistical parallax, it becomes possible to calculate the absolute magnitude of each type of star. When stars of a recognisable type are subsequently identified much further away, it then becomes possible to determine their distances by measuring their apparent magnitude. Methods of distance determination using such methods are discussed in chapter 9, and can be applied to any single 'key' star provided it is visible through an optical telescope.

Apparent and absolute magnitude are discussed in the next section.

THE APPARENT MAGNITUDE SCALE

The apparent magnitude scale provides a means of classifying stars according to their brightness, as they appear to an observer on earth.

The 'brightness' of a star is determined by the flux of light energy received at the earth from the star.

The first scale of stellar magnitude was devised by Hipparchus of Babylon in 127 B.C. and refined by Ptolemy of Alexandria in 137 A.D., and still forms the basis of the present day classification system.

From a subjective evaluation Ptolemy classified the stars into a scale of 6 visual magnitudes, with the brightest stars having magnitude 1 and the faintest stars having magnitude 6.

In 1856 Pogson accurately determined that a first magnitude star is visually 100 times brighter than a sixth magnitude star, and so was able to quantify the magnitude scale.

There is a difference of 5 magnitudes between a first and a sixth magnitude star, and since 5 magnitudes correspond to an increase in brightness of 100, then one magnitude difference corresponds to an increase in brightness of

$$(100)^{1/5} = 2.512$$

Thus a first magnitude star appears 2.512 times brighter than a second magnitude star, which in turn appears 2.512 times brighter than a third magnitude star, and so on.

Also, a first magnitude star which appears (2.512×2.512) times brighter than a third magnitude star is $(2.512 \times 2.512 \times 2.512)$ times brighter than a fourth magnitude star, and so on.

The apparent magnitude scale has been extended to number 24 to include the faintest stars which are just detectable with the largest optical telescopes.

The apparent magnitude scale has also been extended to include negative values for very bright objects, so that the sun, for example, has an apparent magnitude of −26.72, and Sirius, the brightest star, has an apparent magnitude of −1.6.

In order to calculate the difference in brightness between two stars of given apparent magnitude, it is convenient to use the formula derived below.

THE APPARENT MAGNITUDE FORMULA

Consider two stars which have apparent magnitudes m_d and m_b corresponding to light flux measurements of l_d and l_b. The suffixes d and b represent the dimmer star and brighter star respectively. Note that the dimmer star has a larger magnitude number. The ratio of their brightness, l_b/l_d, corresponds to a magnitude difference of $m_d - m_b$, and since a magnitude difference of 1 represents a ratio of brightness of $(100)^{1/5}$, then the magnitude difference of $(m_d - m_b)$ represents a ratio of brightness of $100^{(m_d - m_b)/5}$, that is

$$\frac{l_b}{l_d} = (100)^{(m_d - m_b)/5}$$

Taking logarithms of both sides of this equation we have

$$\log(l_b/l_d) = \frac{(m_d - m_b)}{5} \times \log 100$$

$$= \frac{2}{5} (m_d - m_b)$$

Re-arranging we have

$$m_d - m_b = 2.5 \log(l_b/l_d)$$

(Note: Do not confuse the 2.5 in this formula with the 2.512 brightness factor discussed earlier. They are not the same thing.)

Worked Examples

1. Calculate the brightness ratio between a first magnitude star and a fourth magnitude star.
   We have

$$l_1/l_4 = 100(4 - 1)/5$$

   therefore

$$l_1/l_4 = 100^{3/5} = (2.512)^3 = 15.85$$

2. A certain Cepheid variable star shows variations in apparent visual magnitude between 3.7 and 4.3. Calculate the ratio of brightness for the star at its brightest and least bright stages.
   Using

$$\log(l_b/l_d) = \frac{1}{2.5} (m_d - m_b)$$

   we have

$$\log(l_b/l_d) = 0.4(4.3 - 3.7) = 0.24,$$

$$l_b/l_d = \text{antilog } (0.24) = 1.738$$

THE ABSOLUTE MAGNITUDE SCALE

Stars may differ in apparent magnitude either because they possess different luminosities (see page 23) or because they are situated at different distances from earth. The 'absolute' magnitude scale classifies stars according to the brightness they would exhibit if they were each situated at the same distance from earth. The distance upon which the scale is based has arbitrarily been chosen as 10 pc.

The apparent brightness of a star is related to its distance from an observer by the inverse square law of radiation; that is, the intensity of radiation varies inversely as the square of the distance from the source. For example, if a star was removed to twice its usual distance from an observer, then its apparent brightness would decrease by a factor of 4.

If l represents the apparent brightness of a star at its actual distance from earth, and l_{10} represents the apparent brightness which the star would have if it were situated at a distance of 10 pc from earth, then we have from the inverse square law of radiation

$$\left(\frac{l_{10}}{l}\right) = \left(\frac{d}{10}\right)^2$$

We can use this relationship to find the absolute magnitude (M) of a star if we know the apparent magnitude (m). Using

$$m - M = 2.5 \log\left(\frac{l_{10}}{l}\right)$$

we have

$$m - M = 2.5 \log\left(\frac{d}{10}\right)^2 = 5 \log\left(\frac{d}{10}\right)$$

therefore

$$M = m - 5 \log\left(\frac{d}{10}\right)$$

This formula enables the absolute magnitude to be calculated from the apparent magnitude, provided that the distance to the star is known.

Expanding further we have

$$M = m - (5 \log d - 5 \log 10)$$

therefore

$$M = m - 5 \log d + 5$$

If the distance to the star is given in terms of its parallax angle (p) in arc seconds we can substitue for $1/p''$ for d, which gives

$$M = m + 5 \log(p'') + 5$$

This formula enables the absolute magnitude to be calculated from the apparent magnitude if the parallax angle of the star is known.

Worked Example

Calculate the absolute magnitude of the sun, given that its distance from the earth is 4.8×10^{-6} pc and its apparent magnitude is -26.7.
Using

$$M = m - 5 \log(d/10)$$

we have

$$M = -26.7 - 5 \log \left(\frac{4.8 \times 10^{-6}}{10} \right)$$

therefore

$$M = -26.7 + 31.6 = 4.9$$

and hence

$$M_\odot = + 4.9$$

Note: The symbol $_\odot$ is used to represent the sun, and the symbol $\star$ is used to represent a star. The absolute magnitudes of the sun and a star may be represented by $M_\odot$ and $M_\star$ respectively.

MAGNITUDE DETERMINATION AT DIFFERENT WAVELENGTHS

We have seen that the eye, the photographic plate and photoelectric photometers have different spectral sensitivities, and also that the intensity of stellar radiation is not uniform across the wavebands but exhibits an intensity distribution which is closely akin to a black body radiation curve. It therefore follows that the magnitude we assign to a star will depend upon some discrimination factor which is a property of the detecting instrument.

Consider two stars which have respective peak intensities at, say, 450 nm and 550 nm. Referring to figure 5.9 we can see that on a photographic plate the first star would appear to have a greater magnitude than the second, whereas visually the second star would appear to have a greater magnitude than the first. This effect results from the photographic plate and the eye having peak sensitivities at different wavelengths.

When magnitudes were first obtained photographically, the only photographic plates available were those which were restricted in sensitivity to the blue region of the spectrum, and the term *photographic magnitude* (M_{pg}) refers to magnitude determinations which were made upon plates with a peak spectral sensitivity of around 420 nm. *Visual magnitudes* (M_v) represent a magnitude scale which is based upon the spectral response curve of the human retina which exhibits a peak sensitivity at about 540 nm.

When panchromatic films became available, it was possible by using filters to obtain a photographic response which was identical to the visual response. *Photovisual magnitudes* (M_{pv}) therefore represent a magnitude scale obtained from plates with a peak sensitivity at 540 nm.

We have seen that the photomultiplier tube is unique among light detectors, in that it has a very broad spectral sensitivity curve. It is equally sensitive to all regions of the visible spectrum, and with this device and a combination of filters it is possible to make magnitude determinations at any wavelength. Also if the glass envelope is replaced by quartz or fluorite and measurements are made from a satellite or rocket, then magnitudes may be measured in the ultra violet.

The U.B.V. magnitude system samples radiation in the ultra violet, blue and visual regions, and the magnitude of a star may be defined in terms of its magnitude at these three wavelengths.

Figure 8.4 shows the transmission characteristics of the filters which are used in the U.B.V. system, for a light source of equal intensity for the whole range of wavelengths shown. The visual curve corresponds to the visual response curve of the human eye.

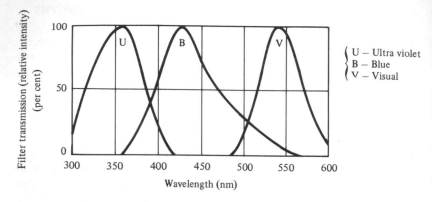

Figure 8.4   *Transmission characteristics of the U.B.V. system filters*

COLOUR INDEX

If the magnitudes of a star are determined in, say, the blue and visual regions respectively, then the difference between the magnitudes at these two different wavelengths is known as the 'colour index' (C.I.), that is

$$C.I. = m_{pg} - m_v = M_{pg} - M_v$$

Note that since we are considering only one star we can use apparent or absolute magnitudes in the formula.

Blue magnitudes are determined photographically using an unsensitised emulsion, which responds only to blue wavelengths. The 'blue' magnitude is therefore represented by the abbreviation m_{pg} where pg stands for 'photographic'.

The usefulness of the colour index is that it provides a means of approximating the temperature of the surface of the star, since it samples the radiation curve of the star at two different places. The colour index identifies, by virtue of the difference in magnitude at the two wavelengths, the shape of a particular black body curve to which the radiation curve of the star corresponds. Once such a curve is identified the temperature of the star is known. This is illustrated in figure 8.5.

The curves show that the colour index ($m_{pg} - m_v$) is negative for a hot star when m_v is a higher number than m_{pg}, and positive for a cooler star when m_v is a lower number than m_{pg}. Note that we are concerned with magnitudes, and the brighter star is assigned the lower magnitude number. The magnitude scale begins with high values of magnitude at the origin,

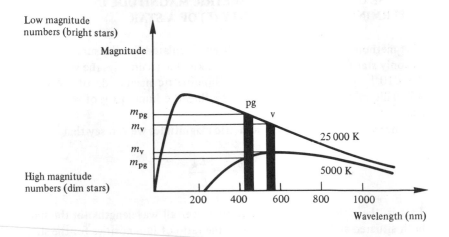

Figure 8.5 *Illustrating how the colour index identifies a stellar radiation curve*

corresponding to 'dim' stars, and these values decrease along the ordinate of the graph, as the stars become brighter.

A colour index of zero is obtained for a star at a temperature of 10 000 K.

BOLOMETRIC MAGNITUDES

The magnitude systems which have been discussed so far, involve only certain regions of the spectrum. However, a photomultiplier tube is sensitive to all regions of the spectrum and when it is operated above the obscuring effects of the earth's atmosphere, say from a satellite, it is possible to measure the total amount of radiative flux arriving from a star at all wavelengths.

The total radiative flux is measured in Watts m^{-2}. Magnitudes based upon such measurements are known as 'bolometric' magnitudes. If the distance to the star is known, then the absolute bolometric magnitude may be determined as before.

Thus a photomultiplier measurement made from above the earth's atmosphere of the 'total' radiative flux arriving from the star can be used to determine the 'apparent' bolometric magnitude. The absolute bolometric magnitude can then be found using:

$$M = m - 5 \log(d/10)$$

THE USE OF ABSOLUTE BOLOMETRIC MAGNITUDE IN DETERMINING THE LUMINOSITY (L) OF A STAR

This method uses data which have been calculated for the sun. The sun is the only star for which the luminosity is known accurately, the value being 3.9×10^{26} Watts. Also the absolute bolometric magnitude of the sun is 4.7. This information can be used to find the luminosity of some other star.

Since we are considering bolometric magnitudes we can say that

$$\frac{l_{10\star}}{l_{10\odot}} = \frac{L_\star}{L_\odot}$$

That is, the ratio of total radiative flux over all wavelengths for the stars, both situated at 10 pc, is the same as the ratio of luminosities for the stars.

Hence the relation

$$M_{\text{bol}_\odot} - M_{\text{bol}_\star} = 2.5 \log\left(\frac{l_{10\star}}{l_{10\odot}}\right)$$

may be rewritten as

$$M_{\text{bol}_\odot} - M_{\text{bol}_\star} = 2.5 \log\left(\frac{L_\star}{L_\odot}\right)$$

Provided that the absolute bolometric magnitude for the star has been previously determined, then the luminosity of the star may be calculated.

Bolometric Corrections

If the magnitude of a star is determined on some arbitrary scale it is possible to convert this to a bolometric magnitude using a bolometric correction B.C., such that for visual magnitudes for example

$$\text{B.C.(visual)} = m_{\text{bol}} - m_{\text{v}} = M_{\text{bol}} - M_{\text{v}}$$

Note that the same bolometric correction factor is applicable to apparent and absolute magnitudes. Bolometric correction factors were once obtained from considerations regarding theoretical stellar models. However, these correction factors have now been confirmed by observations made from satellites operating above the earth's atmosphere, where the total 'bolometric' flux arriving from a star may be measured.

PHOTOMETRIC COMPARISONS

The apparent magnitude of a star may be measured either photographically or by photoelectric detectors. There are advantages and disadvantages in either technique, that is

1. Although both methods are sensitive, the photomultiplier yields the highest percentage accuracy, allowing magnitude determinations to the second decimal place.
2. The response to light intensity of a photomultiplier is linear, and a calibrated instrument may give a direct reading. Obtaining equivalent information from a photographic plate requires extrapolation from the 'H and D' curve, which gives rise to inaccuracies.
3. The disadvantage of the photomultiplier is that it can only be used to analyse one star at a time and, if it were the only device available, then undue demands for telescope time would be made upon the few very large telescopes which are in existence.

In practice, a useful compromise is made between photomultiplier and photographic techniques. A star 'field' is recorded photographically but a number of 'reference' stars within the field have their magnitudes measured independently by photoelectric means. The remaining stars on the plate have their magnitudes determined 'photometrically' by comparison with the reference stars.

Photometric comparison of the amount of blackening of a photographic plate due to star images is now a fully automated process. It involves measuring the extent to which light is prevented from traversing the plate by various densities of silver deposited.

9 The Classification of Stars and Further Methods of Determining Stellar Distances

SPECTRAL CLASSIFICATION

We have seen that stellar spectra are characterised by thousands of absorption lines, with most elements being represented. We have also seen that the temperature of a star affects the distribution of intensity across its continuous spectrum (chapter 2). Whenever a spectrogram of a star is examined these two phenomena are superimposed, and certain Fraunhofer lines appear much stronger than others, depending upon which area of the star's spectral curve is enhanced at the particular temperature of the star. In the case of the sun which has a surface temperature of about 6000 K, CaII* absorption lines are particularly enhanced, as are the lines FeI and some other neutral elements. Molecular bands of CH are also prominent.

The reason for the enhancement of lines is that the atoms concerned exhibit their spectral lines at wavelengths for which the solar radiation intensity happens to be greatest. Helium absorption lines do not show up in the sun's spectrum because they occur at wavelengths which are not radiated by the sun, since it is not hot enough. Helium lines show up only in hotter stars.

As a further example we consider the reasons why hydrogen 'Balmer' lines are particularly enhanced in stars with a surface temperature of 10 000 K. The origin of 'Balmer' lines was discussed in chapter 2.

The presence of an absorption line in a stellar spectrum results from the outer electrons of atoms, situated in the relatively cooler atmospheric layers of the star, absorbing some of the radiant energy from the stellar interior. The electrons involved are capable of absorbing a variety of dis-

---

*Note on spectroscopic notation. In this section, chemical element symbols are sometimes followed by a Roman numeral (for example, CaII). 'I' indicates a neutral atom. 'II' indicates an atom which has lost a single electron, 'III' indicates an atom which has lost two electrons, and so on.

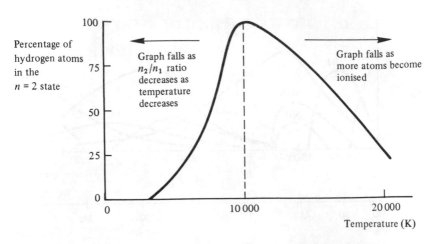

Figure 9.1 *Showing how the percentage of hydrogen atoms in the* n = 2 *state varies with temperature*

crete wavelengths which depend upon the initial and final energy states of the electron. If a 'Balmer' absorption line is to be present in a stellar spectrum, the stellar atmosphere must contain an abundance of 'excited' hydrogen atoms with electrons residing in the $n = 2$ energy level, since Balmer absorption lines result from transitions of electrons from this level. Only at high temperatures could the inter-atomic collisions within the atmosphere involve sufficient energies to excite a significant number of hydrogen atoms to this level.

Although the relative strength of the Balmer lines in stellar spectra increases with temperature there is a limit. At higher temperatures the hydrogen atoms tend to become completely ionised, so reducing the population of hydrogen atoms in the $n = 2$ state, and reducing the intensity of the Balmer lines. Thus there is a steady increase in the intensity of Balmer lines as the temperatures approach 10 000 K and the $n_2 : n_1$ ratio increases, then above 10 000 K there is a decrease in the intensity of the Balmer lines as hydrogen atoms become completely ionised (see figure 9.1).

The relative intensity of Balmer lines in stellar spectra therefore indicates the temperatures of the stars. For example hydrogen Balmer lines are very weak in the spectrum of the sun, since 6000 K is too low a temperature for there to be many electrons in the $n = 2$ state.

The strengths of the absorption lines of other atoms are all subject to the same manner of variation. However, for each type of atom the spectral

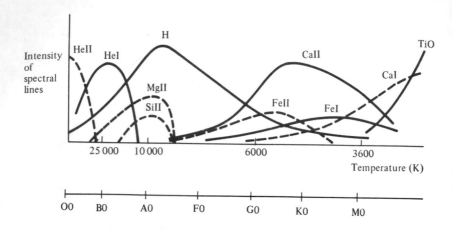

Figure 9.2 *Showing the temperatures at which the absorption lines of some important atoms have their maximum intensity*

line intensity has its maximum value at some other characteristic temperature. The temperature of any star can therefore be found by identifying the atoms which give rise to the most intense spectral lines. Figure 9.2 shows the temperatures at which the absorption lines of some important atoms have their maximum intensity.

It is possible to arrange nearly all stellar spectra in a sequence according to which absorption lines are prominent, and this spectral sequence is also a 'temperature sequence'. A given star may be classified as being a particular 'spectral type' and the 'Henry Draper' or 'Harvard' spectral classification system divides the spectral sequence into seven groups designated by the letters O, B, A, F, G, K and M. Each group is subdivided into ten divisions, for example: G0, G1, G2, . . ., G9, K0, K1, etc. The limits of these groups are shown in figure 9.2, and the temperatures corresponding to the spectral classes can be compared. The sun, for example, is classified as G2.

Figure 9.3a shows some stellar spectra which are typical of groups O, A, G and M respectively. Figure 9.3b extracts the most important features of figure 9.3a. The G band referred to in the spectra is a group of CH and Fe lines, which are prominent in the G and M groups.

It should be noted that the absorption lines of figure 9.3 are predicted by the 'curves' of figure 9.2. A summary of the principal spectral features of the main spectral classes is given in table 9.1.

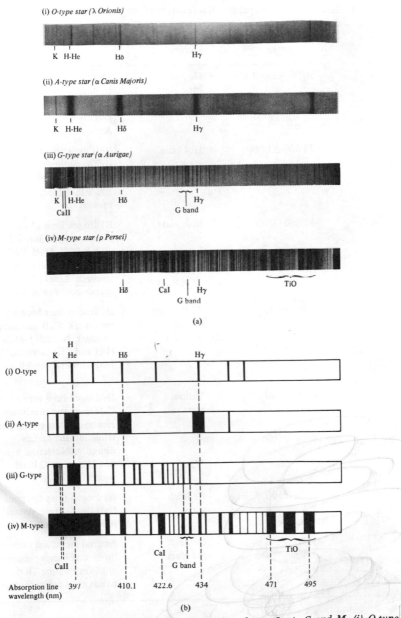

(i) *O-type star (λ Orionis)*

K   H-He   Hδ   Hγ

(ii) *A-type star (α Canis Majoris)*

K   H-He   Hδ   Hγ

(iii) *G-type star (α Aurigae)*

K   H-He   Hδ   Hγ
CaII   G band

(iv) *M-type star (ρ Persei)*

Hδ   CaI   Hγ
G band   TiO

(a)

H
K   He   Hδ   Hγ

(i) O-type

(ii) A-type

(iii) G-type

(iv) M-type

CaII   CaI   TiO
                  G band

Absorption line   397   410.1   422.6   434   471   495
wavelength (nm)

(b)

Figure 9.3 *(a) Stellar spectrograms representing classes O, A, G and M. (i) O-type star (λ Orionis); (ii) A-type star (α Canis Majoris); (iii) G-type star (α Aurigae); (iv) M-type star (ρ Persei) (photographs reproduced by permission of The Royal Astronomical Society). (b) Summarising the detail of (a)*

Table 9.1 Showing the principal spectral features of the various spectral classes

Spectral class	Temperature range	Stellar features	Principal spectral features
O	50 000–25 000	Hottest blue stars	Strong HeII lines. Hydrogen 'Balmer' lines distinct but weak compared with classes B and A
B	25 000–11 000	Hot blue stars	HeII lines absent. Hydrogen lines increase in strength from B0 to B9. SiII, OII and MgII present
A	11 000–7500	Blue stars	Hydrogen lines of maximum strength. Ionised metals (SiII, MgII, FeII, TiII) reach maximum strength. Lines of neutral metals begin to appear
F	7500–6000	White stars	Hydrogen lines become very weak, CaII becoming stronger. Neutral metals (FeI and CrI) as strong as ionised metals by F8 and F9
G	6000–5000	Yellow stars	Hydrogen lines very weak, CaII lines at a maximum. Neutral metals strengthen while ionised metals diminish. Molecular bands of CH appear. G band prominent
K	5000–3500	Reddish stars	Hydrogen lines disappear. CaII lines strong, neutral metal lines strong
M	3500 down	Coolest red stars	Neutral metal lines strong, molecular bands prominent. Very strong TiO bands are a major feature

THE HERTZSPRUNG–RUSSELL (H–R) DIAGRAM

We have seen in chapter 2 that the luminosity of a star is given by the formula

$$L = 4\pi R^2 \sigma T_{\text{eff}}^4$$

which indicates that the luminosity, radius and effective temperature of a star are mutually dependent quantities. The H–R diagram illustrates the interdependence of these three parameters, and is obtained by plotting the positions of stars, on axes representing 'absolute magnitude' and 'spectral class'. Absolute magnitude is a measure of luminosity, and spectral class is a measure of effective temperature. An H–R diagram is shown in figure 9.4 and the effective temperatures associated with each spectral class are included.

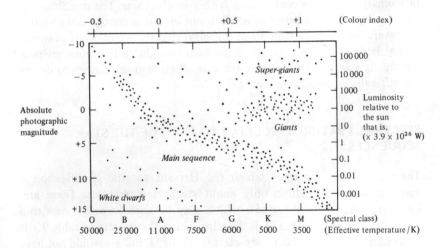

Figure 9.4 *A Hertzsprung–Russell diagram*

The position of a star on an H–R diagram is an indication of the star's radius, since if, say, a star is found in the bottom-left of the diagram by virtue of its having a high temperature and a small absolute magnitude, then it follows that it must be a relatively small star. Conversely, a star plotted in the top-right of the diagram must be very large.

When large numbers of stars are plotted on an H–R diagram, it becomes evident that they fit into certain groupings.

Each grouping represents a particular sequence of stars known respectively as 'white dwarfs', 'main sequence', 'giants' and 'super-giants'.

Ninety-five per cent of all stars, including the sun, belong to the main sequence. The main sequence is a family of stars of chemical composition, similar to the sun, but having different mass. The more massive the star, then the higher is its temperature and consequently its magnitude.

Above the main sequence lies a group of stars known as the 'giants'. They are mainly of spectral types G and K and are on average 10 magnitudes brighter than the main sequence stars of types G and K.

The 'super-giants' are relatively rare. They are up to ten times larger than a giant and about 100 times as luminous. Betelgeuse is an example of a super-giant, being a very bright star in the constellation of Orion. The red colour of the star is just perceptible to the naked eye.

'White dwarfs' on the other hand have very high temperatures but low luminosity, and so we infer that they must have small diameters. White dwarfs are interesting because although they are small, they are unusually dense. Their existence is usually detected when they are one of the partners in a binary system, in conjunction with a brighter star. The star Sirius has a white dwarf as a companion which is not luminous enough to be visible. However, from its action on Sirius it is clear that the white dwarf possesses a very large mass. White dwarfs have been thought to be somewhat rare, but this is now known not to be the case. Their assumed rarity was due to the difficulty of detecting them.

SPECTRAL FEATURES PECULIAR TO EACH OF THE STAR SEQUENCES

The spectral features upon which the Harvard spectral classification is based provide information only about stellar temperatures. There are, however, other spectral features which may be used to assign stars to a particular sequence on the H–R diagram. Using the criteria of table 9.1 it is possible to assign a star to say class G, K, or M, but we would not have enough information to say whether the star was a super-giant, giant or main sequence star. We could, of course, make this decision if we knew the absolute magnitude of the star, but it may be that the star is out of range of the distance-determining methods discussed so far. It is in this situation that the 'other' spectral features are important.

One spectral feature of particular significance is that, for a given temperature, the spectral lines of main sequence stars are stronger than the spectral lines of giants, and these in turn are stronger than the spectral lines of super-giants. The atmosphere of a super-giant is less dense than that of a giant because it is extended, and as a result the degree of ionisation in super-giants is greater. Consequently, there are fewer atoms in a state where they are able to absorb photons and contribute to the Fraunhofer line production. The same phenomenon is observed when the spectral lines of giants are compared with those of main sequence stars. Hence by comparing the relative strengths of the spectral lines, the three groups may be distinguished.

Another important spectral feature which enables white dwarfs to be distinguished from main sequence stars is that the spectral lines of white dwarfs are very broad compared with main sequence stars which are at the same temperature. This broadening arises because the atmospheres of white dwarfs are very dense, and the close proximity of the atoms causes the principal quantum energy levels, within the atoms, to split into many 'sub-levels'. Rather than a discrete wavelength being emitted, a band of wavelengths is emitted, giving rise to a broad spectral line.

STELLAR CLUSTERS

There are six parameters which distinguish any individual star. These are luminosity, effective temperature, radius, mass, chemical composition and age. In preceding chapters we have seen some of the methods by which the first five of these may be determined, but there exists no direct method of assessing the age of a star, and age is a primary consideration in the study of stellar evolution.

However, when the motions of stars within our own galaxy are examined, it is found that many of them do not move independently, as does for example the sun, but move about the galaxy in groups known as 'clusters'. It is a reasonable assumption that stars which are found in clusters have been in association since their birth, and although the age of stars within a cluster cannot be determined, it is reasonable to assume that these stars are all of approximately the same age. With this assumption it becomes possible, on examination of the members of a cluster, to be reasonably sure that any differences between them are not due to age differences. Hence study of a cluster eliminates this unknown parameter.

Two types of stellar cluster have been distinguished, and are designated as 'globular' clusters and 'open' clusters. Open clusters are sometimes referred to as 'galactic' clusters. Globular clusters are the larger, containing between 10^5 and 10^6 stars, whereas the smaller open clusters may contain a few hundred stars. About 200 globular clusters and 1000 open clusters have been identified. Stars in open clusters are fairly evenly spaced and individual stars are distinguishable. In globular clusters there is a considerable concentration of the stars towards the centre, which appears as a single luminous mass. Representations of the two types of cluster are shown in figure 9.5.

Open clusters are often found to contain large quantities of interstellar dust and gas, whereas this is not the case with globular clusters. It is also observed that the brightest members of open clusters are of spectral types O and B, which are high temperature, 'blue' stars, of small radius, whereas the brightest members of globular clusters are relatively cool, 'red' stars of large diameter. The presence of types O and B stars in open clusters indicates that open clusters are relatively young in terms of the cosmic time

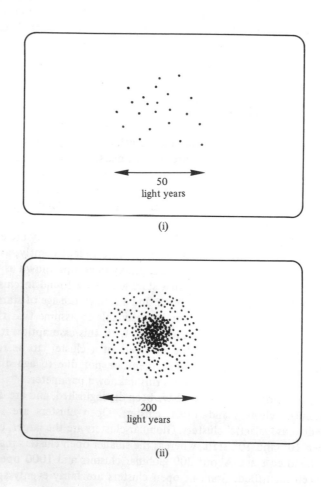

Figure 9.5 *(i) An open cluster; (ii) a globular cluster*

scale, since these particular stars are using their energy reserves at such a prodigious rate, that they can have been in existence only for some 10^7 years. By way of comparison, the sun has existed for some 4.5×10^9 years.

It is also significant that the two types of cluster are characteristic of different parts of the galaxy. Globular clusters are associated with the galactic nucleus, while open clusters are situated in the spiral arms.

STELLAR POPULATIONS

The characteristic differences between the two types of cluster, suggest the possibility that the stars contained in them may represent two different stellar species. This hypothesis is in fact verified when H-R diagrams are constructed for individual clusters. Figure 9.6 shows typical H-R diagrams for an open cluster and a globular cluster.

The H-R diagram for a typical open cluster resembles the general H-R diagram for non-cluster or 'field' stars. However, the main sequence is particularly narrow, and there is a distinct gap between the top of the main sequence and the red giant stars. This is known as the 'Hertzsprung' gap.

The H-R diagram for a typical globular cluster is quite distinct, having a very short main sequence, which curves away to the top-right of the diagram, into the red giant area. There is also a track of stars which runs from the red giant area to the left of the diagram, and with a slight decrease in magnitude. In the middle of this track are the R. R. Lyrae stars, which are a group of 'intrinsic variable' stars which exhibit periodic fluctuations in luminosity.

Since the two types of cluster contain stars which are intrinsically distinct, stars of open clusters and globular clusters are designated as 'population I' and 'population II' stars respectively. However, since this original demarcation, some four intermediate population types have been recognised, with populations I and II being regarded as extremes.

The H-R diagram distinguishes population I from population II stars on the basis of luminosity, effective temperature, and radius. If spectra of the two population types are examined, it is found that there are also important differences in 'chemical composition'. Compared with population II stars, population I stars contain about ten times more of elements heavier than helium, though it must be remembered that even population II stars comprise over 95 per cent hydrogen and helium.

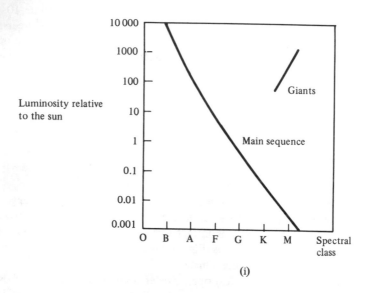

(i)

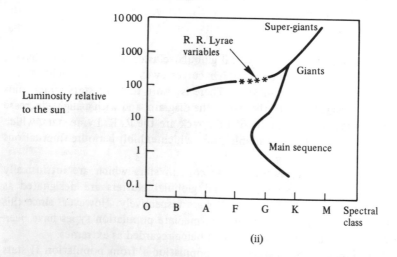

(ii)

Figure 9.6 *(i) H–R diagram – open cluster (population I stars); (ii) H–R diagram – globular cluster (population II stars)*

THE DISTRIBUTION OF POPULATION I AND POPULATION II STARS IN OUR GALAXY

It has been mentioned that open clusters (population I stars) are associated with the spiral arms of the galaxy and that globular clusters (population II stars) are associated with the galactic centre. If 'field' stars, which are any stars not included in a cluster, are examined in terms of their position on an H-R diagram and their chemical composition, it becomes possible to assign them to either population I or population II. We are situated in a spiral arm of the galaxy, and examination of 'local field stars' shows that this spiral arm contains more population I than population II stars. Furthermore, when the motions of these stars are analysed, it is found that the population I stars, including the sun, follow a circular orbit around the galactic centre, whereas the population II stars follow elliptical orbits which take them close to the galactic centre. Hence there is a definite association of population II stars with the galactic nucleus and of population I stars with the spiral arms.

It is also found that the many stars with properties intermediate between populations I and II describe orbits which are elliptical, but which are more rounded than the orbits of population II stars.

THE ORIGINS OF POPULATION I AND POPULATION II STARS

The study of stellar populations shows that our galaxy contains stars of various ages, with the younger stars of population I being associated with the spiral arms, and possessing a higher proportion of the heavier elements.

From this it is surmised that the galaxy originated as a cloud of hydrogen gas, possibly slightly contaminated with heavier elements from exploded stars in earlier galaxies. Gravitational forces caused the cloud to shrink, but also individual clusters of gas to form, from which the first stars or clusters of stars condensed.

The original hydrogen cloud probably revolved slowly, but the speed of revolution increased as the clouds mass concentrated towards its centre, and its moment of inertia decreased. As a consequence of the rotation, the sphere became flattened and more disc shaped. Stars formed in regions where the gas density happened to be highest, and stars of large mass passed quickly through their life histories, producing heavier elements in a series of thermonuclear processes. At the end of their lives these stars exploded and returned their constituent elements, including the new heavier elements, to the inter-stellar medium. This medium became increasingly enriched in the heavier elements, with the result that more recently formed stars contain a higher proportion of them.

The older stars we observe now were formed when the galaxy was young, and deficient in heavier elements; these are population II stars. At the present time, inter-stellar gas and dust is concentrated in the spiral arms of the galaxy, and it is these regions where new stars are being formed. The new stars are enriched in the heavier elements and hence the spiral arms are characterised by population I stars.

Other spiral galaxies exhibit the same distribution of old and young stars. For example, when the Andromeda galaxy is photographed through a red filter, the globular clusters and the red giants of population II are highlighted in the central disc, whereas with a blue filter, the hot, young stars of population I trace out the spiral arms.

THERMONUCLEAR REACTIONS IN STARS

A stable atom invariably has a smaller mass than the combined masses of its constituent particles. For example, the deuterium atom ^2_1H has a mass of 2.014 102 u, where the masses of its constituent proton and neutron are 1.007 825 u and 1.008 665 u respectively (u is the atomic mass unit and has the value 1.66×10^{-27} kg). The total mass of the separated proton and neutron is 2.016 490 u. This 'mass defect' is therefore

$$
\begin{array}{r}
2.016\,490 \text{ u} \\
-\,2.014\,102 \text{ u} \\
\hline
=\,0.002\,388 \text{ u}
\end{array}
$$

Therefore when the proton and neutron combine, 0.002 388 u of matter disappear! Einstein has shown that matter and energy are equivalent entities, and that matter can disappear and re-appear as 'energy'. The relation involved in mass–energy conversion is given by the equation

$$E = mc^2$$

Hence the amount of energy produced when a proton and neutron combine to form a deuterium atom is

$$
\begin{aligned}
E &= 0.002\,388 \times 1.66 \times 10^{-27} \times (3 \times 10^8)^2 \\
&= 3.57 \times 10^{-13} \text{ Joules}
\end{aligned}
$$

or more conveniently

$$= \frac{3.57 \times 10^{-13}}{1.6 \times 10^{-19}} = 2.23 \text{ MeV}$$

The energy equivalent of the mass defect is known as the 'binding energy' of a nucleus. In order to break up the nucleus into its constituent particles, this same amount of energy would need to be supplied by an external source.

Reactions in which lighter nuclei are combined to form heavier nuclei are called 'fusion reactions'. They can occur only at extreme conditions of temperature and density. Fusion reactions produce the thermonuclear energy of the stars.

The fundamental thermonuclear reaction in stars is the fusion, by collision, of hydrogen nuclei to form helium nuclei. This reaction takes place in stellar interiors by either of two distinct series of processes. The first of these is the 'proton–proton chain', in which the collisions of protons results in the direct formation of heavier nuclei, whose subsequent collisions yield helium nuclei. The second is the 'carbon cycle' (or CNO cycle) which consists of a series of stages in which carbon nuclei absorb a succession of protons until eventually they undergo an alpha-decay, whereupon emitting the alpha particle ($_2^4$He), they become carbon nuclei once more.

The *proton-proton chain* proceeds as follows.

The initial reaction is the formation of a deuterium atom by the direct combination of two protons (that is, two hydrogen nuclei)

$$_1^1\text{H} + _1^1\text{H} \longrightarrow _1^2\text{H} + e^+ + \nu$$

where e^+ is a positron and ν is a neutrino.

A deuterium atom may then join with a proton to form a $_2^3$H nucleus

$$_1^1\text{H} + _1^2\text{H} \longrightarrow _2^3\text{He} + \gamma$$

where γ is a gamma ray photon.

Finally, two $_2^3$He nuclei react to produce a helium nucleus and two protons

$$_2^3\text{He} + _2^3\text{He} \longrightarrow _2^4\text{He} + _1^1\text{H} + _1^1\text{H}$$

(Note that the first step in the reaction occurs twice.)

The mass of the helium nucleus plus two positrons is 4.0037 u, while the total mass of the four original protons is 4.0313 u. The mass defect is therefore 0.0276 u and so the energy liberated by the complete reaction is found using $E = mc^2$

$$E = \frac{0.00276 \times 1.66 \times 10^{-27} \times (3 \times 10^8)^2}{1.6 \times 10^{-19}} = 25.7 \text{ MeV}$$

For the 'carbon cycle' we have

$$^1_1\text{H} + ^{12}_6\text{C} \longrightarrow ^{13}_7\text{N}$$

$$^{13}_7\text{N} \longrightarrow ^{13}_6\text{C} + e^+ + \nu$$

$$^1_1\text{H} + ^{13}_6\text{C} \longrightarrow ^{14}_7\text{N} + \gamma$$

$$^1_1\text{H} + ^{14}_7\text{N} \longrightarrow ^{15}_8\text{O} + \gamma$$

$$^{15}_8\text{O} \longrightarrow ^{15}_7\text{N} + e^+ + \nu$$

$$^1_1\text{H} + ^{15}_7\text{N} \longrightarrow ^{12}_6\text{C} + ^4_2\text{He}$$

The net result is the same as for the proton–proton chain, which is the formation of a helium nucleus and two positrons from four initial protons. Again the energy released is 25.7 MeV. The initial $^{12}_6\text{C}$ nucleus acts as a catalyst in the reaction, since it re-appears at the end of the sequence, unchanged. Obviously the carbon cycle can only operate in a star in which appreciable carbon is present, and so it is only found in population I stars.

The proton–proton chain and the carbon cycle are alternative reactions which produce the same end product. In general, the carbon cycle is more efficient at high temperatures and the proton–proton chain is more efficient at lower temperatures. Hot stars with temperatures above 2×10^7 K, obtain their energy largely from the former cycle, while population I stars which are at temperatures less than this, and all population II stars, obtain most of their energy from the latter cycle. In the case of the sun, both processes take place, but the proton–proton chain is the most important.

There is a precedent for the order of nuclear fusion reactions, since lighter nuclei react more readily than heavier nuclei. This is because heavier nuclei possess a greater positive charge and therefore experience larger repulsive forces from each other. Therefore, in a star which contains both hydrogen and helium, it is the hydrogen fusing reaction which continues until all the available hydrogen is used.

It is possible to estimate the 'total energy store' of the sun, at least as far as the reactions of converting hydrogen to helium are concerned. When four protons of total mass 4.0313 u combine to form a helium nucleus and two positrons, the mass defect is 0.0276 u, as we have seen. If we consider the total mass of available hydrogen in the sun, then the fraction 0.0276/ 4.0313 = 0.0068 will be converted to energy. However, only the hydrogen in the central core of the sun, which represents about one-tenth of the total mass of the sun, is at sufficiently high enough temperature and pressure to undergo fusion. The total thermonuclear energy available to the sun is found using $E = mc^2$, therefore

$$E = 0.0068 \times 0.1 \times 1.99 \times 10^{30} \times (3 \times 10^8)^2 = 1.23 \times 10^{44} \text{ J}$$

(where 1.99×10^{30} kg is the mass of the sun).

The luminosity of the sun at the present time is 3.9×10^{26} W and so the conversion of hydrogen to helium in the sun should last for approximately

$$\frac{1.23 \times 10^{44}}{3.9 \times 10^{26}} = 3.2 \times 10^{17} \text{ s} = 10\,000 \text{ million years}$$

The solar system is estimated to be 5000 million years old, and so the sun appears to be roughly halfway through its life, in its present form.

When a star has exhausted its supply of hydrogen, the next lightest element, helium, undergoes fusion. The core of the star undergoes a collapse before this reaction can occur so that the core temperature increases from around 10^7 K to around 10^8 K. The increased temperature of the collapsed core causes the outer parts of the star to expand, and stars at this stage in their evolution undergo a dramatic increase in radius.

The helium fusion reaction is not straightforward. When two helium nuclei interact by collision they form a beryllium nucleus, but this nucleus is highly unstable and splits up again immediately. In order for a heavier element to be formed, it requires that a beryllium nucleus of a particular resonant state be formed, which is then able to capture a third alpha-particle to produce the element 'carbon'. This is the *'triple-alpha reaction'*, and is the reaction which occurs in stars.

At higher temperatures still, further fusion reactions can take place; for example, helium reacts with carbon to produce oxygen, or with oxygen to form neon, or with neon to form magnesium, and so on.

With such reactions it is possible to start with hydrogen and produce any of the elements. However, when elements heavier than iron are formed the reactions require heat to be supplied, rather than giving heat out. Such 'endothermic' reactions occur at the moment when an ageing star of large mass explodes as a 'supernova'.

139

STELLAR EVOLUTION

From the previous account it is clear that stars evolve through a number of stages, from the original condensation stage when the first hydrogen fusion reaction begins, through a series of other fusion reactions which can drastically alter the stars temperature and radius, and then on to some final stage when the supply of nuclear fuel is exhausted. Changes in radius, effective temperature and luminosity during this evolution change the position of the star on an H-R diagram, to the extent that the history of the star can be mapped as a complex pathway.

A 'proto-star' is a condensing cloud of dust and gas, mostly hydrogen, which has not yet reached high enough temperature and pressure for a hydrogen fusion reaction to begin. Proto-stars, however, are strong emitters of infrared radiation. Because they are cooler than stars and of very low luminosity, they are plotted in the bottom-right of an H-R diagram. During contraction, gravitational potential energy lost by inward falling atoms is converted to kinetic energy, and the temperature therfore increases. At high enough temperatures, hydrogen and helium atoms (if present) become ionised. When the core temperature reaches about 10^7 K, a hydrogen fusion reaction begins.

The time taken for this contraction to occur depends upon the amount of material present; the more massive the proto-star the faster it contracts; for example, a proto-star of equivalent mass to the sun takes some 10^6 years to condense into a star, whereas a proto-star with half of the sun's mass takes over 10^8 years.

The mass of a proto-star also determines its path across an H-R diagram until fusion commences, and the new star then appears in the main sequence. Figure 9.7 shows the tracks on an H-R diagram of three proto-stars of different masses, as they approach the main sequence. The pathways are not straight because of changes which occur in the mechanism of heat transfer within the star, which is a balance between convection and radiation.

The star maintains a steady position on the main sequence for as long as the hydrogen fusion reaction continues. It has been mentioned that whether the proton–proton chain or the carbon cycle operates depends upon temperature. It is the more massive stars which become hot enough for the carbon cycle to operate. The main sequence is therefore subdivided into the 'upper' and 'lower' main sequence according to which type of reaction predominates. The more massive stars are found higher in the main sequence. See figure 9.8.

The length of time for which a star remains in the equilibrium state on the main sequence depends upon the total mass of hydrogen in its core,

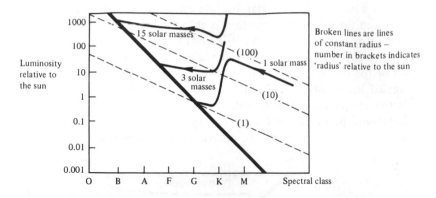

Figure 9.7 *Tracks formed on an H–R diagram by three proto-stars of different masses*

and the rate of hydrogen consumption. Both factors are determined by the total mass of the star. The time spent on the main sequence for stars of various masses is shown in figure 9.8.

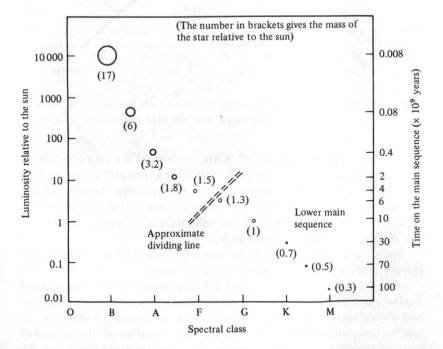

Figure 9.8 *Showing the upper and lower main sequence*

When a main sequence star has almost exhausted its hydrogen fuel supply and the triple-alpha reaction is about to begin, large changes in the structure of the star occur and it leaves the main sequence. The core of the star collapses and the resulting increase in temperature causes the outer parts of the star to expand. The increase in core temperature results in an increased luminosity, but the increase in radius causes a decrease in the effective temperature. The star follows a complicated track across the H–R diagram and becomes a red giant or super-giant. A simplified version of some of the pathways for stars of different mass is shown in figure 9.9.

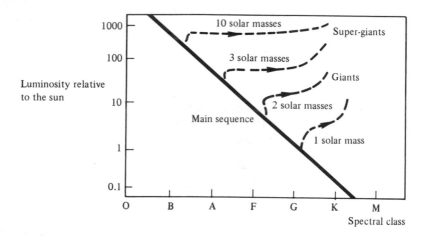

Figure 9.9 *Some evolutionary pathways for stars of different masses in which the hydrogen fuel supply is almost exhausted*

At a temperature of about 10^8 K the triple-alpha reaction commences, and at this stage the star turns back, following a new path across the H–R diagram towards the main sequence, and finally down through the low luminosity regions. In stars of 'low' mass the onset of helium fusion is a violent reaction and is known as the 'helium flash'. In stars of higher mass the onset of helium fusion is less violent.

Evolutionary tracks in the latter stages of a star's life are complex, but the death of a star occurs in one of just a few ways.

In the case of a star of mass of 1.4 solar masses or less, the series of nuclear reactions runs its course until these reactions become endothermic, and during this rundown the star becomes a white dwarf. Eventually the star becomes non-luminous and is known as a 'black dwarf'. In the case of stars of very low mass, the sequence of nuclear reactions may not reach the endothermic stage, since low mass gives an insufficiently high central temperature for the later exothermic reactions to proceed.

For stars of mass in excess of 1.4 solar masses, the final gravitational collapse occurs with sufficient force to form a *neutron star*. The collapse produces a rebound shock wave which ejects the outer shell of a star at high velocity. The kinetic energy of the ejected material is transferred to the surrounding inter-stellar gas, and raises the temperature of the gas to millions of Kelvins. This shell of hot gas appears as a *nebula*. Simultaneously, enormous amounts of radiation are emitted from the neutron star which is at the centre, which may briefly appear as a *supernova* and may have a luminosity equivalent to an entire galaxy!

While the outer shell of the star is 'blown' off, the inner core of the star continues to collapse. The pressure and density of the core increases to the extent where protons and electrons combine together to form neutrons. Further increase in pressure causes the individual nuclei to split up into their constituent neutrons and the few remaining protons. Finally the core reaches a state where further compression is impossible and the core stabilises as a dense neutron gas from which the star derives its name.

The radius of a neutron star is relatively minute at about 10 km, which gives a mean density of about 10^{17} kg m^{-3}.

Some neutron stars are magnetic and may be rotating rapidly, and under these circumstances the radiations produced at all wavelengths, from gamma rays to radio waves, are given off in short, regular bursts. Such neutron stars are known as *pulsars*. The periods of pulsars range between 0.03 to 4 seconds, and the duration of the pulses ranges between 20 and 50 milliseconds. The Crab Nebula in the constellation Taurus is the remnant dust cloud of a supernova explosion, which was recorded in 1054 A.D. The neutron star at the centre of this cloud is in fact a pulsar. This pulsar has been observed at radio, visible, X-ray and γ-ray wavelengths, and has a pulse period of 33 milliseconds. The energy radiated by a pulsar derives from the rotational kinetic energy of the star and, as energy is given out, the period of rotation decreases in time, with a corresponding decrease in the pulse period.

An alternative fate for a star of mass greater than 1.4 solar masses is that it may become a *black hole*, although no black hole has been observed directly. Theoretical models show that if a massive star undergoes a 'sudden' collapse, the explosion producing the supernova may not occur, and the resulting density of material may approach infinity. Such an object is known as a black hole since the gravitational attraction of the body prevents even the escape of radiation! The 'Einstein Observatory', which is an astronomical satellite discussed in chapter 11, has produced observational evidence for the existence of a black hole. This is discussed on page 172.

FURTHER METHODS OF DETERMINING STELLAR DISTANCES

(i) Distance Determination by Spectroscopic Parallax

The ordinate of the H-R diagram shown in figure 9.4 is the absolute photographic magnitude, but this quantity can only be obtained if the distance to the star is known. It is therefore only possible in the first instance to plot on an H-R diagram those stars which are near enough to have their distances determined by trigonometrical parallax. However, once this has been done, and when it is clear that the stars fit into sequences, it then becomes possible to use the H-R diagram itself, to determine the distances to stars outside the range of trigonometric parallax. In fact the distance to any star may be found provided that it is bright enough to give a spectrogram.

The method involves first classifying the star on the Harvard scale using the criteria of table 9.1. Then by identifying 'other' spectral features, the star is classified into a particular 'sequence'. Once the star's spectral class is known and also to which sequence it belongs, it is possible to trace across, on the H-R diagram, to find the absolute photographic magnitude of the star. Once the absolute photographic magnitude of the star is known it is then a matter of using the formula

$$m - M = 5 \log(d/10)$$

to determine the distance to the star. The value of m is obtained using a photoelectric photometer.

As an example, consider figure 9.10.

Suppose that in the spectrogram of a star, spectral lines are present which indicate the spectral class as being A0, and in addition suppose that the spectral lines are sharp, so that we know we are dealing with a main sequence star rather than a white dwarf. Referring to figure 9.10, we can trace upwards from class A0, until we reach the main sequence stars. We can then trace across to find the value of absolute photographic magnitude, which in this case is about −2.

From figure 9.10 it is clear that the method of spectroscopic parallax is not too precise, because of the vertical scatter of the main sequence stars. The error in determining the absolute photographic magnitude may be ±1 magnitude, which in turn gives a 50 per cent error in the final evaluation of distance to the star! While such an error would be unacceptable for most scientific work, because of the vast distances involved such distance determinations are still significant.

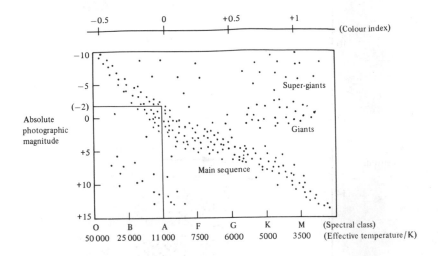

Figure 9.10 *Illustrating spectroscopic parallax*

(ii) Distance Determination by 'Main Sequence Fitting'

This method is quite accurate and is applicable to clusters of stars. If a plot of 'apparent' magnitude versus spectral class is made for stars in a single cluster, then this can be considered as being an H-R diagram for the cluster, since the distance between the stars is small compared with their distance from the earth. That is, the relative brightnesses of the stars give a true representation of their relative luminosities. Figure 9.11 shows a plot of apparent magnitude versus spectral class for some stars in the globular cluster M 67.

In this diagram the main sequence and the turn-off point to the giant region are distinct. The actual technique involves sketching out the cluster diagram on semi-transparent paper, and superimposing this on a calibrated H-R diagram which has 'absolute' magnitude as ordinate. The cluster diagram is then moved until the main sequences of both diagrams coincide, at the 'same' spectral types. If the ordinate values of magnitude have the same scale, it is possible to read the difference $(m - M)$ for any star, which represents the difference between the star's apparent magnitude and absolute magnitude. This quantity may then be substituted into the formula

$$m - M = 5 \log(d/10)$$

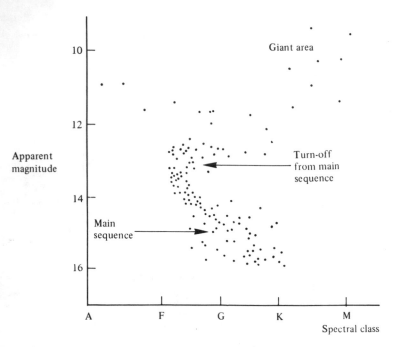

Figure 9.11 *A plot of apparent magnitude versus spectral class for some stars in the globular cluster M 67*

and the distance to the star obtained. The accuracy of the method lies in the fact that $(m - M)$ may be found for a large number of stars, and the average value for d may be calculated.

(iii) The 'Cepheid Variable' Method of Distance Determination

Many stars show periodic fluctuations in luminosity. The fluctuations may be due to the star being part of an eclipsing binary system, but there are, in addition, many single stars which vary in luminosity on their own account. These are known as 'intrinsic variables'.

Various categories of intrinsic variables exist, but the Cepheid variables are a very important type because they provide a means of distance determination, which extends as far as distant galaxies. One well known Cepheid is Polaris.

The significant feature of Cepheid variables is that a relation exists between their periods of light variation and their absolute magnitudes at median light. (Median light is the time midway between minimum and maximum values of brightness.) The brighter Cepheids have longer periods, and a Cepheid with a period which is ten times longer than another Cepheid is approximately 2 magnitudes brighter at median light.

The relation between period and brightness for Cepheids was discovered by Henrietta Leavitt of Harvard University in 1912, who observed Cepheids in the Magellanic clouds, which are two small, companion galaxies to our own. Light curves for twenty-five Cepheids were plotted in the small Magellanic cloud, but Cepheids in the large Magellanic cloud were not found as useful, since the light from these stars is masked by inter-stellar dust in the galaxy.

Further work has shown that there are two types of Cepheids, each having a different period–brightness relationship. The 'classical' Cepheids are the more luminous and are population I stars, belonging to galactic spiral arms. The 'W. Virginis' Cepheids are population II stars and are often found in globular clusters. The relation between period and absolute visual magnitude for both types of Cepheids is shown in figure 9.12.

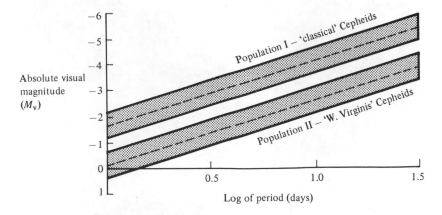

Figure 9.12   *The relation between period and absolute visual magnitude for the two types of Cepheid variable stars*

In 1912 the exact distances to the Magellanic clouds were not known, and there are no closer Cepheids which are in the range of trigonometrical parallax. However, the Magellanic clouds represent star 'clusters', such that the distances to all of the stars in one particular cluster can be taken as being approximately the same. That is, the distance across each cluster is negligible compared with the distance of the cluster from earth. Under these circumstances it became possible to determine the distances to the Magellanic clouds, using the method of main sequence fitting.

A knowledge of the distance to the Cepheids in the small Magellanic cloud enabled their absolute magnitudes to be determined using

$$m - M = 5 \log(d/10)$$

Cepheids are all giants or super-giants, and are intrinsically very luminous, to the extent that they show up in distant galaxies. It is, therefore, possible to measure the period of light variation for such a distant Cepheid, and from this derive the absolute magnitude at median light. The apparent magnitude of the Cepheid is measured with a photoelectric photometer, and the distance to the Cepheid, and the galaxy which contains it, can be evaluated by again using the formula

$$m - M = 5 \log(d/10)$$

10 Radio Astronomy

The earth's atmosphere is opaque to almost the entire electromagnetic spectrum, with the exception of visible and radio wavelengths, and so we are provided with an optical 'window' and a radio 'window' through which to view space. In this chapter we examine radio astronomy in detail.

THE ORIGINS OF RADIO ASTRONOMY

Radio waves from space were first detected in 1935 by K. G. Jansky, an American radio technician who at the time was experimenting with radio receivers which were 'directional'. Janksy discovered radio waves which came from the direction of the Milky Way, and which reached maximum intensity when the antenna was pointed to the constellation of Sagittarius, which is the direction of the galactic centre. Radio technicians refer to these radio transmissions as 'galactic noise'.

Many varied sources of radio waves have since been identified in space, and the importance of radio astronomy is that it provides a different, but complementary picture to optical observations. There are also certain phenomena which are only detected at all by radio observation. The emission of radiation from the corona of the sun is an example.

Radio waves are generated by charged particles, especially electrons, undergoing acceleration. Hence a fully ionised gas like the corona of the sun, which emits no visible radiation because there are no electron transitions between quantum energy states in atoms, is still a strong emitter of radio waves.

RADIO AND OPTICAL TELESCOPES COMPARED

The only significant difference between visible and radio radiation is the range of wavelengths involved, and consequently radio and optical telescopes are essentially of the same design. Both are constructed to capture as much radiation as possible, and to focus it on to a 'detector'. The detector for visible radiation may be the human retina, a photographic plate or a photoemissive surface, but for radio waves the detector is a dipole antenna. A dipole antenna is a metal rod in which electric currents are caused to oscillate at the same frequency as the incoming radiation, and it is the length of the dipole which determines the range of wavelength to which the radio telescope will respond. The radio telescope does not form 'images' and the incoming radio signals are transduced into electrical currents by the dipole. Figure 10.1 illustrates how similar in principle are the design of a radio telescope and an optical reflecting telescope.

Radio waves are reflected from electrically conducting surfaces in the same manner as visible radiation is reflected from a mirror. The reflecting dish of a radio telescope may consist only of a wire mesh surface built upon a simple framework. Provided that the spaces within the wire mesh are small compared with the incoming wavelengths, the fact that the reflecting surface is not continuous is of no significance. For microwaves of wavelength less than 10 cm, sheet metal must be used in place of wire mesh.

THE RESOLVING POWER OF OPTICAL AND RADIO TELESCOPES COMPARED

The resolving power of a telescope is a measure of the ability of the telescope to distinguish fine detail. The resolving power of a telescope increases with the diameter of the parabolic reflecting mirror in the case of a reflecting telescope, or with the diameter of the objective lens in the case of a refracting telescope. Resolving power decreases if the wavelength of radiation is increased.

According to a formula derived by Lord Rayleigh, the smallest angular separation of two points which can just be resolved by a telescope is given by

$$\alpha_0 = 1.22 \, \frac{\lambda}{D}$$

where α_0 is the angle in radians subtended at the objective lens of the telescope by the two points, λ is the wavelength of radiation in metres, and D is the diameter of the 'collecting' mirror or lens in metres. *Note:* a small value for α_0 means a high value of resolution.

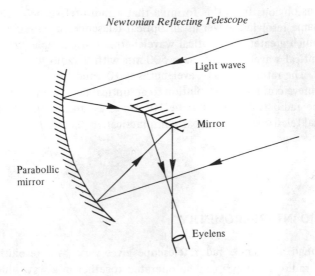

Newtonian Reflecting Telescope

Light waves

Mirror

Parabollic
mirror

Eyelens

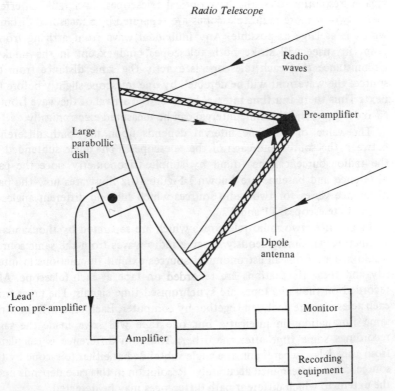

Radio Telescope

Radio
waves

Pre-amplifier

Large
parabollic
dish

Dipole
antenna

'Lead'
from pre-amplifier

Monitor

Amplifier

Recording
equipment

Figure 10.1 *Comparison of the Newtonian reflecting telescope and a radio telescope*

It is obvious from this formula that a radio telescope could never give the same resolving power as an optical telescope, since radio wavelengths are much greater than optical wavelengths. For example, we can compare an optical wavelength of, say, 500 nm with a radio wavelength of, say, 5 cm. The ratio of these wavelengths is 10^5 and, if a radio telescope were to achieve comparable resolution to an optical telescope, then the diameter of the radio telescope must be 10^5 times as large as the diameter of the optical telescope. This is clearly impractical.

RADIO INTERFEROMETRY

Although any single radio telescope gives very poor resolution, two or more radio telescopes can be operated together in a way which gives far greater resolution than the best optical telescopes. In a radio interferometer two or more radio telescopes are separated by a measured distance, which is as large as possible. Any individual wave front arriving from a point in space will strike both telescopes, and except in the unlikely circumstance that each telescope is exactly the same distance from the source, the wavefront will be detected by one telescope slightly before the next. Thus there is a time interval between the arrival of the wave front at each detector, and this time interval can be measured electronically.

The value of the time interval depends upon the 'path difference' between the source and each of the telescopes. (The angle subtended by the radio source can be found by simple trigonometry once the path difference and baseline are known.) Figure 10.2 illustrates how the path difference varies for two radio sources which subtend different angles at the radio telescopes, T1 and T2.

In practice, two radio telescopes which are separated by thousands of kilometres can simultaneously observe radio waves from the same source. Radio signals from all astronomical sources exhibit fluctuations in intensity and these fluctuations are recorded on tape at each telescope. Also recorded on the same tapes are synchronised time signals. The tapes from each telescope are analysed together by computer. Each tape will show the same fluctuations in intensity, but one tape will have made the same recordings some time after the other. The path difference is calculated from this delay, whereupon the angle subtended at either telescope by the source can be determined accurately. Resolution in this case depends upon the extent to which different path differences may be detected.

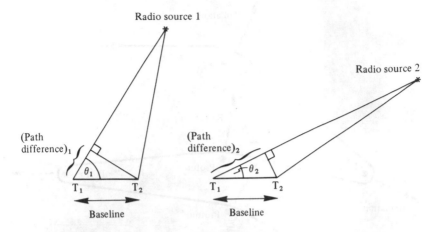

Figure 10.2 *Illustrating how the path difference varies for two radio sources which subtend different angles at the radio telescopes T1 and T2*

THE ORIGINS OF RADIO WAVES IN SPACE

There are two 'main' processes in which electrons are accelerated within radio sources.

Thermal radiation is produced when free-moving high-speed electrons come under the influence of the electrostatic field of a positive ion, in which case they are constrained to follow a hyperbolic path, radiating as they do so. Figure 10.3 shows a free electron travelling at high speed in a hot gas cloud. During each interaction with a proton, the electron is accelerated and emits a burst of radiation at radio frequency. Thermal radiation is the main source of radio wave production in the hot gases of the solar atmosphere.

Synchrotron radiation is emitted wherever fast-moving electrons interact with a magnetic field. Under the influence of the magnetic field the electrons are constrained to move along 'helical' paths and emit radiation at radio frequencies as they do so. Synchrotron radiation is responsible for the production of radio waves in the corona of the sun.

The production of synchrotron radiation is illustrated in figure 10.4.

RADIO EMISSION FROM THE SUN

The sun is a continuous source of intense radio emission. The intensity of

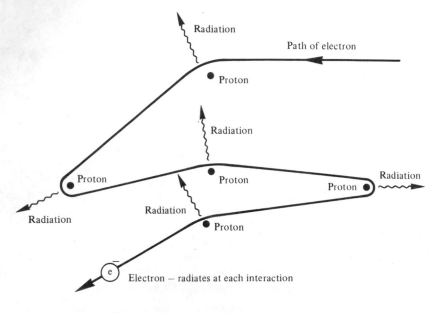

Figure 10.3 *The mechanism of thermal radiation*

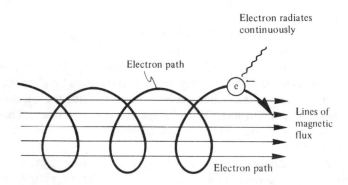

Figure 10.4 *The mechanism of synchrotron radiation*

the emissions is observed to increase when there is an increase in sunspot activity within the photosphere. Both the radio emission and the sunspot activity fluctuate periodically with an 11-year cycle. The radio wavelengths associated with sunspot activity are typically in the range of about 10 cm.

In addition to the radiation associated with sunspot activity, the occurrence of solar flares from the chromosphere results in sudden, strong increases in the sun's long-wave radiation at wavelengths of about 55 cm. These bursts of radiation may last for several minutes and have intensities which are thousands of times greater than the continuous background radiation of the sun. The earliest radio emissions that were detected from the sun were due to solar flares, and were noticed by British radar technicians in 1942 as a result of the considerable disturbance they caused to radar transmission. The disturbances were known as *solar noise*.

Radio emissions from the sun with wavelengths up to 1 metre are found to be radiated from a surface area which coincides with the sun's visible disc. However, radiation at wavelengths greater than 1 metre is also radiated from the corona.

Wavelengths of 5 metres from the corona are particularly strong. The 5 metre radiation is synchrotron radiation which emanates from the high densities of free ions and electrons which are trapped in and accelerated by the sun's magnetic field.

RADIO EMISSION FROM THE PLANETS

All of the planets emit some radiation at radio wavelengths, typically around 3 cm, but this radiation is always very weak and difficult to detect against background radiation.

Jupiter is of particular interest since from time to time it emits radio waves at wavelengths between 13 and 15 metres and also between 7 and 11 metres. These radiations are as yet unexplained but they do appear to originate from a particular place on the surface. Among the hypotheses put forward to explain this radiation are that it emanates from lightning flashes, or it is due to radiation from ions which have been ejected by the sun and subsequently caught up and accelerated in Jupiter's magnetic field.

RADIO EMISSION FROM OUR GALAXY AND BEYOND

Radio emissions from our galaxy exhibit a continuous spectrum of wavelengths between 10 cm and 15 cm. These emissions are known as 'galactic noise'. The intensity of the galactic noise roughly equates with the distribution of stars in the galaxy.

Apart from the general background of galactic noise there are certain discrete sources of intense radio emission. A few thousand of these sources have been catalogued but the nature of the sources has been established only in about half the cases. Examples of objects which have been identified as radio sources are listed below.

1. The remnant gas clouds of supernovae (exploding stars) are strong radio sources, emitting wavelengths of about one metre. The Crab Nebula in the constellation of Taurus is one of the most intense discrete radio sources.

2. Gas clouds or 'nebulae' other than those associated with supernovae are composed mostly of hydrogen, and radiate at wavelengths of 21 cm which is the characteristic radio wavelength of hydrogen gas. However, if the nebulae contain within them, or are in close proximity to, a very hot star (for example class O, B or A), then the temperature of the gas cloud may rise to 10 000 K. Ionisation which results, produces free ions and electrons, and interactions between these give rise to thermal radiation at wavelengths around 10 cm. Examples include the 'large' nebula in Orion, the 'Horse Head' nebula in Orion, and the 'North America' nebula in Cygnus.

3. Galaxies other than our own are radio emitters. There are two categories

   (i) 'Normal' galaxies such as the Andromeda galaxy and the Magellanic clouds radiate their own galactic noise. The main emission from the Andromeda galaxy has a wavelength of 73 cm. In some spiral galaxies the radiation arises from the nucleus of the galaxy, whereas in other galaxies the radiation arises from a region which extends beyond the visual galaxy, from a region known as the 'galactic halo'.

   (ii) 'Radio' galaxies are very strong emitters of radio waves, but in all cases they are visually very faint, and it has proved difficult to observe them. It would seem that in many cases each source is a 'double' galaxy. In some instances the two galaxies involved are separate, but in others the two galaxies appear to be either in collision or else are a single galaxy which is dividing in two. Cygnus A is an example of a double galaxy system where the two individual galaxies merge.

   An interesting feature of most radio galaxies is that there are two centres of radio emission which are separated by considerable distances on opposite sides of the 'optical' galaxy. This is illustrated in figure 10.5.

The separation between the centres of radio emission and the central optical galaxy may range between 15 and 500 kpc. This phenomenon is thought to be the result of violent explosions occurring within the galaxy sending out streams of charged particles in two opposite directions. If these charged particles impinge on a gas cloud which extends throughout

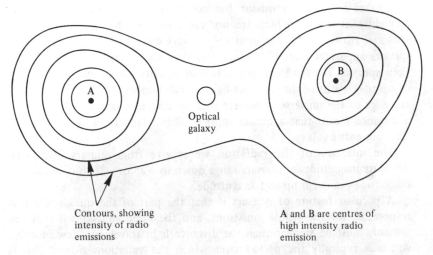

Contours, showing
intensity of radio
emissions

A and B are centres of
high intensity radio
emission

Figure 10.5 *Intensity contours of a radio galaxy*

or beyond the galaxy, then the magnetic fields which are always associated
with such gas clouds cause the charged particles to spiral and emit synchrotron radiation. This is illustrated in figure 10.6.

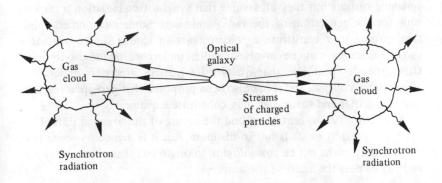

Figure 10.6 *Synchrotron radiation from gas clouds surrounding a galaxy*

QUASARS

The word 'quasar' is a contraction of 'quasi-stellar radio source' and came
into use when it was believed that all quasars were strong radio sources.

The exact nature of a quasar has not been determined, but it is now thought that quasars which are not radio emitters far outnumber those that are. However, the non-emitters are very difficult to detect. Optically, quasars appear as faint blue stars, but they show very large red shifts in their spectra. The red shifts indicate that quasars have recession velocities of up to 9/10ths of the speed of light, which implies that they are close to the edge of the universe! The fact that we can see quasars at all over such a distance means that a quasar must be about 100 times more luminous than our entire galaxy.

The intensity of the radiation we receive from quasars fluctuates. Absolute magnitudes of quasars range down to -25 or -26, but changes in magnitude can occur up to 1 magnitude.

A peculiar feature of quasars is that the part of the quasar which is responsible for the visible emissions, and the radio emissions if they are present, must be smaller than the distance light travels in a few months, which is typically the period over which the variations occur. This is deduced from the short periods over which the intensity variations occur. If the radiating area was larger, then the intensity variations would last longer.

This reasoning indicates that a quasar is very small in astronomical terms and in fact is about 10^{17} times smaller than our own galaxy.

The small size of a quasar requires that there must be an enormous concentration of energy in a very small region. A number of theories surround quasars but they all assume that synchrotron radiation is responsible for the generation of the radio emissions. Some astronomers think that quasars may constitute a 'chain' reaction among supernovae, or a single supernova may be involved which undergoes periodic outbursts. Other theories postulate that quasars are massive superstars undergoing gravitational collapse, and it has also been proposed that perhaps a reaction between matter and antimatter may constitute a quasar.

A controversy also centres around the nature of the large red shift. That the large red shift exists is not in disupute, but it is argued by some that the red shift might not be cosmological in origin and that quasars may not be situated near the limits of the universe.

21 cm RADIO WAVES

In certain regions of inter-stellar space there are abundant quantities of hydrogen gas. The temperature of inter-stellar hydrogen is very low, between 50 and 100 Kelvin, and consequently the hydrogen atoms are not ionised and their electrons are in the ground state.

So far, we have considered the ground state of the hydrogen atom as representing a 'single' energy level. We calculate the value of this energy level by considering the electrostatic potential energy which the electron possesses at the particular orbital radius. However, there is another factor which is to be considered; the proton and electron in the atoms each have a 'spin' motion, and spinning charges generate magnetic fields. The magnetic potential energy of the electron must be considered in addition to its electrostatic potential energy.

The electron may spin in the same direction as the proton, which gives a certain value for the magnetic potential energy of the electron or the electron may spin in the opposite direction to the proton which gives a lower value for the magnetic potential energy. There are therefore two sub-energy levels of the ground state of the hydrogen atom, and which ever one the electron is occupying depends upon its direction of spin. This is illustrated in figure 10.7.

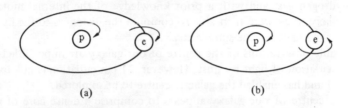

(a)   (b)

Figure 10.7 *The two possible directions of spin for the electron of a hydrogen atom: (a) electron spins in same direction as proton, giving a higher sub-energy level; (b) electron spins in opposite direction to proton, giving a lower sub-energy level*

The two sub-energy levels are known as 'hyperfine' energy states and a spontaneous transition from the higher to the lower of these states can occur with the electron making a sudden change in the direction of its spin. When the transition occurs a quantum of very low energy, equal to the difference in energy between the two states, is emitted. The frequency associated with the quantum is 1420.4 MHz, and the corresponding wavelength is 21 cm.

The hyperfine energy states of the electron may change whenever the hydrogen atoms collide. When collisions take place the hydrogen atoms usually exchange their electrons. If the newly acquired electron has the same spin as the electron which has been lost, then there is no change in the energy state, but if the electrons involved have opposite spins, then there is a change in the energy state, either up or down. A collision may therefore result in excitation, de-excitation or no change.

The density of the gas is so low that each hydrogen atom is involved in a collision, typically once every 400 years. The spontaneous downward transition between the hyperfine energy states occurs only after an average time delay of 11 million years, and that the 21 cm line can be detected at all under these circumstances gives an impression of how vast the inter-stellar clouds of hydrogen gas are.

In addition to being able to detect the presence of hydrogen gas by means of 21 cm wave radiation, it is also possible to detect Doppler shifts from this wavelength and thereby determine any motion of the gas clouds.

21 cm radio wave investigations have been particularly important in determining the structure of our own galaxy. Neutral hydrogen gas is one of the most abundant elements present in the spiral arms of our galaxy, and studies of the 21 cm radiation 'emitted' by the spiral arms has enabled them to be accurately mapped. This mapping was carried out by Van de Hulst, A. B. Muller, and J. Oort of Leiden University. From Doppler shifts of the 21 cm radiation, they measured the radial velocities of the clouds of hydrogen gas, and with a prior knowledge of the internal motions of our galaxy, they were then able to compute the distances to the hydrogen clouds, and hence to the spiral arms.

Visible observations of the centre of our galaxy are impossible because of obscuration of light by dust. However, 21 cm radiation is not absorbed by dust and has enabled the galactic centre to be explored.

The centre of our galaxy appears to comprise a dense core of ionised hydrogen, which is surrounded by a rotating mass of neutral hydrogen which radiates at the 21 cm wavelength. The core of ionised hydrogen is identified by virtue of its absorbing radiation at characteristic wavelengths of 3.5 and 15.2 metres. The speed of rotation of the neutral hydrogen is deduced from Doppler shifts in the 21 cm radiation. The rotating mass of neutral hydrogen is itself surrounded by an expanding mass of neutral hydrogen, which leads into the spiral arms.

RADIO WAVELENGTHS CHARACTERISTIC OF OTHER SUBSTANCES

Numerous other substances, apart from hydrogen, and including some which are organic, have been identified in inter-stellar gas by virtue of their radio emissions. Table 10.1 lists some of these molecules with their charac-teristic wavelengths.

Table 10.1 Characteristic wavelengths of some chemical compounds

Molecule	Chemical formula	Characteristic wavelength (cm)
Hydroxyl	OH	2.2, 3.7, 5.0, 6.3, 18
Carbon monoxide	CO	2.7
Cyanogen	CN	2.6
Carbon monosulphide	CS	0.2
Water	H_2O	1.35
Ammonia	NH_3	1.2, 1.3
Hydrogen sulphide	H_2S	0.18
Hydrogen cyanide	HCN	0.35
Thioformaldehyde	HCHS	9.5
Formaldehyde	HCHO	0.2, 1.0, 2.1, 6.2, 6.6
Methanol	CH_3OH	1.2, 36
Formic acid	HCOOH	18.3
Acetaldehyde	CH_3HCO	28.1

All of the wavelengths quoted in the table penetrate inter-stellar dust, and these radiations are important because they allow investigations into dense gas and dust clouds.

RADAR ASTRONOMY

Radio astronomy involves the analysis of radio waves which originate in space. In 'radar' astronomy a radio transmitter on earth is the source of the radio waves. The radar or 'microwave' beam is directed to objects within the solar system such as planets, or meteors, and the reflected waves are detected by small diameter radio telescopes on earth or on board terrestrial satellites. Subsequent analysis of the reflected waves provides information about the objects.

Radar was developed by British scientists during the Second World War, when it was used to detect German aircraft on bombing raids, as they flew towards Britain over the North Sea and English Channel. Reflected radar waves were used to ascertain the position and numbers of aircraft. The word 'radar' is formed from the initial letters of a radio detection and ranging.

For military and scientific use, the advantage of radar is that the transmissions can penetrate clouds, fog and atmospheric dust, since microwavelengths are significantly larger than any atmospheric particles. Furthermore, the background radiation of the sun contains insignificant microwave radiation and so radar can be used day or night without any solar interference, although microwave emissions from solar flares can interfere with radar transmissions.

The lower wavelength limit for radar transmissions is set by the ability of the waves to penetrate clouds and fog, but there is also an upper wavelength limit, since resolution is inversely proportional to wavelength (see page 150). Also, as the wavelength approaches the same order of magnitude as the object under investigation, the waves are diffracted around the object rather than being reflected, which also makes them unsuitable. Optimum wavelengths are selected that are long enough to penetrate clouds, yet short enough to provide sufficiently good resolution and to detect objects of reasonable size. (Wavelengths between a few centimetres up to one hundred metres are suitable.)

THE USE OF RADAR TO STUDY THE DISTANCE, SURFACE AND ROTATION OF THE PLANETS

The distance over which radar can be operated is limited, because of the loss of energy of the radar beam which tends to spread out as it progresses. The reflected waves are also depleted in energy since some energy is absorbed at the planet's surface, and since planets are spherical much of the incident radiation is not reflected back in the direction of the observer. Thus, the reflection of a radar beam has only been achieved with the nearer planets, namely Mercury, Venus and Mars. Reflections have also been obtained from the moon and the sun. The distance to each of these bodies is accurately determined by timing the interval between the transmitted and reflected pulses. In the case of the sun, reflection of the radar pulse occurs within the corona.

It is possible to use a very narrow radar beam provided that the receiving dish of the radio telescope is very large, and under these circumstances radar maps of the moon, Mercury, Venus, and Mars have been made. The radar maps of the moon are of as high a standard as those made from optical observations. Radar maps of the planets are much less detailed, although in the case of Venus, radar provides the only means of observing the planet's surface, since the dense atmosphere of Venus is opaque to visible radiation. It has been found that the surface of Venus reflects about twice as well as the surface of the moon, which indicates that the surface of Venus is very smooth. Some large geographical features have, however, been identified on the surface of Venus, and one is thought to be a long mountain chain, over a mile high. Radar measurements on the surface of Mars have shown that the height difference between the highest mountains and deepest valleys is about twelve kilometres. Thus the 'roughness' of the martian surface is comparable to that of earth.

The speed of rotation of the planets can be measured from Doppler shifts in reflected radar waves. A radar beam of a given frequency is emitted, but the reflected beam consists of a band of frequencies which range either side of the original frequency. Waves reflected from the advancing limb of the planet show an increase in frequency while waves reflected from the receding limb show a decrease in frequency. The shift in wavelength $\Delta\lambda$ for either the approaching or receding limb is found, and then the tangential velocity v of that limb is calculated from the Doppler formula, which was derived in chapter 7, that is

$$v = \frac{c\,\Delta\lambda}{\lambda}$$

The radius r of the planet can be found by measuring the angle it subtends at the earth, if the distance to the planet is already known. The period of rotation of the planet can be found using

$$\text{period} = \frac{2\pi r}{v}$$

Radar values for the periods of rotation of Mercury and Venus are particularly interesting. In the case of Mercury a radar measurement in 1965 showed the period to be 59 days, where prior to this the period had been taken to be 88 days, measured by visual observations. In the case of Venus no visual measurement could be attempted because of the dense atmosphere of the planet. The period of rotation is found to be 243 days, but unexpectedly it was also found that the rotation of Venus is 'retrograde', in that it rotates in the opposite direction to the other planets.

THE USE OF RADAR IN THE STUDY OF METEORS

Any 'stray' rock fragment travelling within the solar system is known as a 'meteoroid'. Some 25 million meteoroids enter the earth's atmosphere every 24 hours. The majority of meteoroids have speeds between 10 and 80 km s^{-1}, and when a meteoroid enters the earth's atmosphere, the heat generated by friction causes the particle to 'blaze' brightly. A meteoroid which has entered the earth's atmosphere, and which has become visible, is known as a 'meteor'. Most meteors are vaporised completely by the heat before they near the ground, but a small proportion do collide with the earth's surface. A meteor which reaches the ground is known as a 'meteorite'. Damage due to meteorites is very rare, although a few very large meteorite craters are known.

Meteors can be studied photographically. To measure both the direction and velocity of a meteor requires that the meteor is simultaneously photographed by two cameras which are separated by many kilometres. Measurements taken from the two photographs enable the height and direction of the meteor to be calculated. Each camera has a constant frequency 'interrupter' which only allows light to enter the aperture intermittently. On the photograph the meteor appears as a dashed, rather than a continuous line, and the velocity of the meteor can be calculated from the length of the dashes in the trail.

The same type of observations can be made using radar, but with the advantage that observations can be made during daylight hours, and they are not hindered by cloud. The meteors themselves are too small to be detected by radar, but the intense heat they produce leaves behind them a trail of ionised air which does reflect radar waves.

The original radar studies were begun in 1945 by James Hey at Malvern. Sir Bernard Lovell has also studied meteors using the Jodrell Bank radio telescope, and his studies have shown that most, if not all, meteoroids originate from within the solar system. This fact is established by accurately determining the velocities of many meteors. All meteors studied had velocities consistent with them following an 'elliptical' path around the sun. If meteoroids had strayed into the solar system from elsewhere, some might possess velocities too high to follow elliptical orbits, and would instead be taking a 'hyperboloid' path through the solar system.

In radar observations, the distance to the meteor, or its 'range', is determined by measuring the time delay between the emitted and reflected pulses. The direction of the meteor, relative to the radar station, is found by measuring the rate at which the range changes during the flight of the meteor. The height, velocity and direction of the meteor can be found by simultaneous observations from two radar stations.

11 Observations from Beyond the Earth's Atmosphere

INFRARED OBSERVATIONS

The earth's atmosphere is 'largely' opaque to infrared radiation and observations at these wavelengths are only possible if carried out from above the atmosphere.

Observations have been made from high-altitude balloons and rockets, but artificial satellite observations have provided the greatest quantity of data.

A number of satellites have been employed to detect infrared sources, and the most advanced of these is illustrated in figure 11.1. This satellite was known as 'I.R.A.S.' (infrared astronomical satellite) and was launched in January 1983.

I.R.A.S. was essentially a modified Cassegrain telescope. A large concave primary mirror collected radiation which it focused on to infrared detectors, after intermediate reflection from a small secondary mirror. There were a total of 62 individual detectors, and their composition and manner of deployment enabled observations to be made in four separate wavebands between 8 and 119 μm.

Infrared radiation is emitted by any object which has a temperature greater than a few degrees above absolute zero, and this includes the satellites themselves. Consequently, in infrared observations, 'noise' generated by the satellite poses a problem. ('Noise' is any radiation produced by extraneous sources which masks radiation from the source under investigation.) In order to detect very faint infrared sources the satellites must be kept as cold as possible, and in the case of I.R.A.S. the infrared detectors were maintained at a temperature of 2 K by a tank of superfluid helium which surrounded the telescope. The working life of such a telescope is limited by the 'run down' time of the cooling system, and in the case of I.R.A.S. this was 250 days.

Astronomical satellites are always computer controlled, and the I.R.A.S. satellite contained a computer which controlled the satellite and stored observational data. These data were in turn relayed to a 12 metre steerable

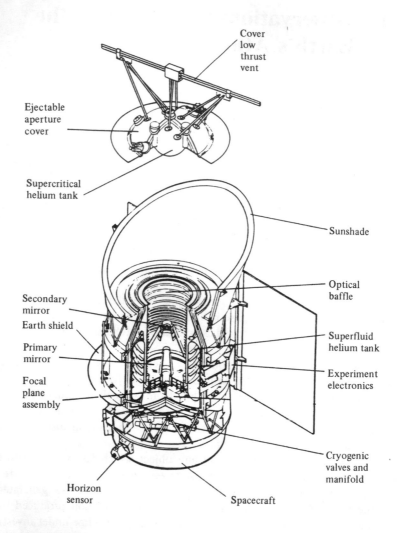

Cover
low
thrust
vent

Ejectable
aperture
cover

Supercritical
helium tank

Sunshade

Optical
baffle

Secondary
mirror

Earth shield

Primary
mirror

Focal
plane
assembly

Superfluid
helium tank

Experiment
electronics

Horizon
sensor

Spacecraft

Cryogenic
valves and
manifold

Figure 11.1 *The I.R.A.S. satellite (reproduced with permission of the Rutherford-Appleton Laboratory)*

dish radio receiver at the control centre at the Rutherford–Appleton Laboratory in Oxfordshire.

Most astronomical objects which can be detected at optical wavelengths have temperatures in excess of 3000 K, and the importance of infrared observations is that objects at lower temperatures, down to tens of Kelvins, may be investigated.

In addition to the investigation of 'cool' objects, infrared observations allow investigation of hot stars which are hidden behind clouds of dust. Visible radiation from such stars is absorbed by dust, since visible wavelengths are of the same order of magnitude as the dust particles. However, the energy is reradiated from the dust at infrared wavelengths, which are much longer, and is not re-absorbed.

Also because infrared radiation penetrates dust, infrared investigations are used alongside radio-investigations to explore the centre of our galaxy.

Infrared astronomy is also important in the investigation of star formation within large dust clouds. Energy is released in the form of infrared radiation as the dust clouds undergo gravitational collapse. It is as the clouds are collapsing into proto-stars that infrared investigation yields the most valuable information.

During the life cycles of stars elements are synthesised, and a highly evolved star may eject these synthesised elements in a supernova outburst, to form a dust cloud around itself. Light from the star is then absorbed by the dust cloud and reradiated as infrared radiation, which on analysis provides information as to the type and quantity of elements involved.

ULTRA VIOLET OBSERVATIONS

There have been two important ultra violet satellite observatories. The first was the National Aeronautics and Space Administration (N.A.S.A.) satellite, known as Copernicus, which was shut down in 1982 after operating successfully for 8 years. The second is the International Ultraviolet Explorer (I.U.E.) which was launched in 1978 and is still operational. The I.U.E. is a joint project of N.A.S.A., the U.K. Science Research Council (S.R.C.) and the European Space Agency (E.S.A.) The I.U.E. has been placed in geosynchronous orbit over the Atlantic Ocean and is operated for 16 hours per day from the United States and 8 hours per day from Europe. The function of these satellites has been the production of both high and low resolution, ultra violet spectra of sources other than the sun. Absorption lines superimposed on these spectra reveal the constitution of inter-stellar gas. The Internal Ultraviolet Explorer is shown in figure 11.2.

The satellite contains a telescope which protrudes at the top, and which is shaded against the sun. A system of baffles inside the telescope tube reduces diffracted sunlight, and any stray light from the earth or moon, to negligible intensity. The apogee boost motor was used to insert the spacecraft into its correct orbit and the hydrazine auxiliary propulsion system was used for manoeuvres during the mission. A star-tracking device

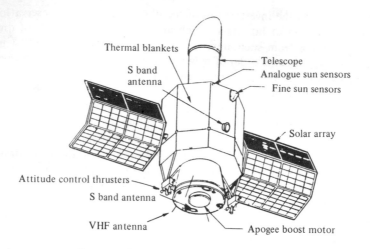

Thermal blankets

S band
antenna

Telescope

Analogue sun sensors

Fine sun sensors

Solar array

Attitude control thrusters

S band antenna

VHF antenna

Apogee boost motor

Figure 11.2 *The International Ultraviolet Explorer (reproduced with permission of the National Aeronautics and Space Administration)*

provides an inertial reference system, which gives the telescope a pointing accuracy of one arc-second.

The telescope has a Cassegrain configuration of mirrors, and the telescope materials were chosen for their light weight, moderate cost, and thermal and optical properties. Beryllium was selected for the primary mirror and fused silica for the secondary mirror. The telescope is thermally insulated and this allows for a nearly constant focus over a wide range of solar aspect angles. Without this insulation, solar radiation could cause differential expansion of parts of the telescope. Small corrections to the focus are made by heaters situated behind the primary and secondary mirrors.

Behind the aperture plate of the telescope are a high-resolution spectrometer and a low-resolution spectrometer. The first can resolve spectral lines with a separation of the order of 0.02 nm, and is suited to the study of the atmospheric characteristics of bright stars and planets. The second has a resolution of about 0.6 nm, and is used in the study of faint sources such as quasars.

The spectrometer detection unit is an 'ultra violet to visible' image converter coupled to a television camera. The video signal is digitised by an analogue/digital converter, and in turn these data are signalled to the ground station.

OBSERVATIONS OF THE I.U.E.

The I.U.E. has been used extensively in the study of hot stars, cool stars, the inter-stellar medium, X-ray objects, extra-gallactic objects and solar system objects.

Two classes of hot stars have been studied. The first class includes hot, sub-luminous stars, which as a group represent the pathway to the end-points of evolution for the majority of stars. The second class includes hot, massive stars in our galaxy.

The study of cool stars has provided information on the characteristics of aged stars. In this case emission line ultra violet spectra are studied which originate in chromospheres, coronae, and any circum-stellar envelopes of gas and dust, and provide information regarding any mass loss due to ageing.

The wide spectral coverage of the I.U.E. of between 115 nm and 320 nm embraces many of the resonance lines of common elements, and the high sensitivity of the detectors makes the I.U.E. a unique instrument for studying the inter-stellar medium. It was previously assumed that the distribution of gas and dust in the inter-stellar medium was simple, but the I.U.E. has shown that this is not the case. A large range of densities and temperatures have been discovered, from cool high-density regions embedded in giant molecular clouds, through rarefied mixtures of ionised and neutral hydrogen, to hot diffuse matter evident only by its X-ray emission and ultra violet absorption. It has been found further that the hot regions are energised by supernovae explosions, and the cool dense regions are characterised by shock waves and ionisation fronts associated with newly formed stars.

The I.U.E. studies have provided information concerning binary X-ray sources. Previously these objects have been inaccessible to ultra violet spectroscopic observations because of their faintness. The I.U.E. findings are helping to discern the physical conditions and dynamics of the material in the sources, and also the interaction between the source and its surrounding medium.

X-RAY OBSERVATIONS

X-ray radiation produced in space is unable to penetrate the earth's atmosphere, and investigation of this radiation is also carried out from rockets and satellites.

Gases heated to temperatures between 10^6 and 10^9 K emit most of their energy as 'thermal' X-rays. In addition, X-rays are produced by the synchrotron radiation process which occurs wherever high-speed charged particles are accelerated in strong magnetic fields. This process was explained in chapter 10 when it was considered as a source of radio waves.

ROCKET FLIGHTS

X-rays were first detected in space in 1948, by a rocket-borne instrument which recorded X-ray emission from the sun. This success led to a programme in which the sun's X-ray output was monitored over an 11 year sunspot cycle. In 1962, after the development of a detector with a sensitivity of 100 times that originally used to investigate the sun, rocket investigation detected the first X-ray source outside the solar system. This was a star in the constellation of Scorpio. Subsequent rocket investigations revealed some 30 other X-ray sources in our galaxy, and an X-ray source which is itself the giant elliptical galaxy M 87.

Information from rocket flights has been sparse because of the limited observation times involved. However, the equipping of satellites as X-ray observatories has put X-ray astronomy on a par with optical and radio-astronomy, with regard to the quantity and detail of the information which is available.

X-RAY SATELLITES

The first X-ray satellite, 'Uhuru', was launched by N.A.S.A. in 1970, and this in turn was followed by a number of others, including a Dutch satellite 'A.N.S.' and a British satellite 'Ariel V'. In 1977, 'HEAO-1' was launched and was the first of an improved group of X-ray satellites known as the 'High Energy Astronomical Observatory' series. Numerous important findings were made by these satellite investigations. However, the second satellite of this series, 'HEAO-2' (also known as the 'Einstein Observatory'), possessed two important innovations which put this satellite into a class apart. These were an X-ray 'telescope' and X-ray television cameras of high sensitivity. The Einstein Observatory was launched in 1978 and remained in orbit for 2 years.

Up to and including the 'HEAO-1' flight, the basic X-ray detector had remained unchanged. It was a version of a Geiger–Müller tube with a thin window which blocked ultra violet radiation. The disadvantages of this simple instrument were that it was not directional and also that it was affected by any cosmic-ray background noise. Fortunately the earth's magnetic field deflects cosmic rays towards the poles, and these detectors could be used provided the satellite was put into an equatorial orbit. The problem of determining the direction of an X-ray source was solved by using complex arrangements of baffles which limited the angle from which incoming radiation was received. Very high angular resolution was possible, but only at high cost in terms of sensitivity.

However, the X-ray telescope could focus X-rays, which were admitted on to highly polished paraboloid and hyperboloid surfaces at minute grazing angles, as shown in figure 11.3.

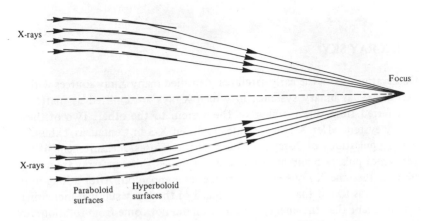

Figure 11.3 *The focusing of X-rays*

The focal length of the telescope was 3.4 m and the aperture diameter was 0.58 m. A conventional concave reflector will not focus X-rays, since apart from the case of minute grazing angles, the X-rays are absorbed rather than reflected. Reflection of X-rays occurs only if the angle between the X-ray and the reflector is minute, and if the reflecting surface is highly polished. Focusing lenses are also not feasible since they do not transmit X-rays.

Two X-ray television cameras were used in the Einstein Observatory. One of these was a 'photon-counting' device capable of detecting X-ray images at the full resolution of the telescope which was 4 arc-seconds, and over a field of 20 arc-minutes. The position and time of arrival of each detected photon was transmitted to the ground where it was stored in a computer. The data accumulated over time intervals ranging from minutes to a few hours were translated into X-ray images of the sky on a television screen. The second imaging X-ray camera utilised a gas proportional counter, which in addition to measuring the position and arrival time of the photons also measured the approximate energy of each photon.

The Einstein Observatory carried a cryogenically cooled spectrometer of medium resolution which was also used to determine the energy of incoming photons, as well as a Bragg crystal spectrometer which had high spectral resolution but with corresponding low sensitivity. The various detectors were arranged on a turntable so that each could be rotated into position at the principal focus of the telescope when required. The arrangement is shown in figure 11.4.

THE X-RAY SKY

Observations from the early satellites identified many X-ray sources within our galaxy as binary systems, in which X-rays are emitted as matter is transferred from one member of the system to the other. Two of these binary systems, Her X-1 in Hercules and Cen X-3 in Centaurus, exhibited regular pulsations of X-rays. In both of these cases, measurement of the periods of pulsation and measurement of Doppler shifts in X-ray frequency showed that the X-ray star was circling its companion. After a long-term study it was found that the orbital speed of the X-ray star was increasing, which meant that the energy for emission did not come from rotation, for then rotational kinetic energy would be lost and orbital speed would decrease. For this reason it is supposed that the source of energy for the X-ray emission is gravitational energy released by the accretion of material from the companion star on to the X-ray emitting object.

Her X-1 and Cen X-3 were later identified visually as a result of improved locations supplied by Uhuru. The combination of X-ray and optical measurements made it possible to infer the mass of the emitting object and, from the change in speed of rotation, the moment of inertia was calculated. The results were consistent with the object being a 'neutron' star with a mass equal to that of one or two solar masses. This was the first measurement of the mass of a neutron star.

Another binary system, Cyg X-1, was found to contain a compact object that flickered irregularly with time intervals of the order of milliseconds. After subsequent optical identification of the system the mass of the X-ray object was estimated to be more than 6 solar masses. Since the object has a large mass but is very compact, it is considered that the object may be a 'black hole'. The X-rays which are produced in the binary system Cyg X-1 emanate from charged particles which are accelerated towards the black hole from the companion star.

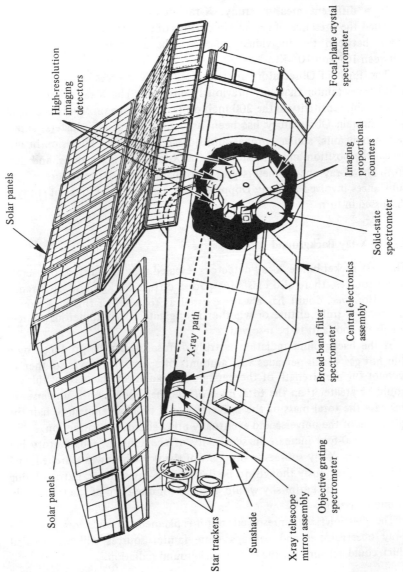

High-resolution imaging detectors

Focal-plane crystal spectrometer

Solar panels

Imaging proportional counters

Solid-state spectrometer

Solar panels

Central electronics assembly

X-ray path

Broad-band filter spectrometer

Star trackers

X-ray telescope mirror assembly

Sunshade

Objective grating spectrometer

Figure 11.4 *The Einstein Observatory (reproduced with permission of the National Aeronautics and Space Administration)*

173

Following these early discoveries astronomers have concluded that 'mass-exchange' binary systems are the source of many and perhaps all of the 'high-luminosity' X-ray sources.

In a different area of study, X-ray observations from Uhuru have revealed the presence of gas within clusters of galaxies, which pervades the space between the individual galaxies. The gas ranges in temperature between 10^6 and 10^9 K.

The Einstein Observatory has detected X-ray sources which are up to 1000 times fainter than any previously observed. The X-ray telescope has comparable sensitivity to the 200 inch optical telescope on Mount Palomar. The Einstein Observatory has been used to study X-ray clusters, supernovae remnants, stellar coronas and quasars. The X-ray television images reveal the position and intensities of the X-ray emitting regions, and data from the X-ray spectrometers give information about the nature of the substances involved and any Doppler shifts. Each of these areas of study is discussed in turn.

(i) The X-ray Background

The first rocket-borne X-ray detectors revealed that in addition to discrete X-ray sources, there also exists in space a diffuse background of X-radiation. However, doubt has always existed as to whether this background radiation is truly diffuse or whether it originates from numerous sources which are too faint to be resolved.

If the background radiation is truly diffuse, it may be produced by a thin hot gas which pervades vast amounts of space. If this was so, then to account for the strength of the background radiation, the mass of this gas would be greater than the total mass of all other bodies in the universe. In this case the total mass of the universe would be large enough to halt the expansion of the universe and eventually bring about its contraction.

The 1000-fold increase in sensitivity of the Einstein Observatory has revealed new X-ray sources in one part of the sky which was searched, and which was previously thought to be devoid of X-ray sources. Extrapolating this result to the entire sky would account for one-third of the total background radiation.

The question is still unresolved, but it is planned to build more powerful X-ray observatories and perhaps more fainter sources will be revealed which could account for the entire background radiation.

(ii) Extra-galactic X-ray Sources

X-ray observations from the Einstein Observatory have principally been centred upon the neighbouring spiral galaxy, M 31 in Andromeda, which is at a distance of two-million light years. Previous X-ray observations of this galaxy showed it to be a faint blur of X-ray emission. The Einstein Observatory has revealed some 80 individual star systems which are comparable to the brighter X-ray sources in our own galaxy.

There is a clear distinction between the X-ray sources in the spiral arms of M 31 and those at the centre. The sources at the centre of the galaxy are assumed to be 'low mass' binary systems created by 'capture' processes which are enhanced by the increased density of stars in this region. The spiral arms are rich in gas and dust, and are regions where new stars are being created. These conditions also seem to favour the creation of very large binary systems that eventually evolve into X-ray binaries of the mass-transfer type. As more information of this type is discovered, the evolutionary processes at work within galaxies should become clearer.

(iii) Clusters of Galaxies

Galaxies tend to cluster in groups, sometimes of hundreds or thousands contained within a distance of a few million light years. Uhuru showed that the space within such clusters is pervaded by gas at temperatures up to 10 K, but the Einstein Observatory has revealed information about the distribution of this gas. In some clusters the gas is concentrated at the centre and decreases smoothly in density with increasing distance from the centre. In other clusters the gas is distributed in 'clumps'.

It is assumed that the gas originates in individual galaxies and that where the gas occurs in clumps within a cluster, this represents a 'young' cluster in which the gas has not had sufficient time to distribute. The older clusters are those with the high concentration of gas at the centre with a smooth density gradation outwards.

(iv) Supernova Remnants

X-ray observations provide information about the collapsed core, the expanding supernova shell, and also how the stellar debris mixes with and heats the inter-stellar gas.

One supernova remnant examined by the Einstein Observatory is 'Cas A' in Casseopeia. The X-ray image of Cas A reveals detail corresponding to the various physical processes occurring in various parts of the remnant. Numerous very bright areas correspond to fast-moving balls of hot gas which are also observable at visible wavelengths. Spectroscopic analysis of these areas indicates that they are enriched in oxygen and sulphur, which is presumed to have been formed by nuclear processes in the interior of the original star. Some X-rays appear to originate in stationary regions which may contain material shed by the star before the explosion and which have now been heated by the shock front. A faint shell surrounding the X-ray images is thought to represent the shock front itself.

The age of Cas A has been determined by measuring the radial speed of the debris, and then by extrapolation, finding the time when all the debris was concentrated at one place. The results give the data of the explosion as being in the second half of the 17th Century.

Theories relating to the mechanisms of supernovae explosions predict the likely occurrence of a pulsar at the centre of supernovae remnants (see chapter 9). However, no pulsar has been identified in Cas A, and since this supernova is relatively young so that any such pulsar has not had sufficient time to move far from the centre, it implies that the theories must be revised. Either no neutron star was created or it has cooled at a faster rate than is allowed for in current models.

(v) Stellar Coronas

The Einstein Observatory has detected X-ray emission from most stars, so that the X-ray emission is now regarded as the norm. Previously, X-ray emission had only been detected from a few types of star, such as white dwarfs, X-ray binaries and cataclysmic variable dwarf novae. Where single stars emit X-rays (for example, a white dwarf) the X-rays are emitted by the stellar coronas, but most theories relating to the heating of stellar coronas predicted that only stars within a temperature range between 5500 K and 10 000 K (that is, types G, F, and A) would emit X-rays at levels comparable to or higher than the sun. The new X-ray observations have therefore had an important influence on this particular area of study. The traditional theories had supposed that energy released by the core which travelled outwards by the processes of convection and radiation is acoustically very noisy, and that this acoustical energy heated the corona. Such theories could only be applied to type G, F, and A stars, since only these types exhibit strong enough convection currents.

Some solar theorists had already postulated that magnetic fields may in some way heat the coronas, but any theories now proposed must fit with the new X-ray observations.

(vi) Quasars

X-ray detectors prior to those in the Einstein Observatory had been able to identify only three quasars near to the earth, having red shifts up to 0.2. The increased sensitivity of the Einstein Observatory has enabled recordings of X-ray emission to be made from every quasar which has been identified optically, including the farthest known quasar with a red shift of 3.5. In addition, many more X-ray sources have been identified which had not previously been identified optically. These objects are in turn now under optical examination so that their red shifts might be examined to determine if they are quasars.

The relatively small size of quasars is inferred from the observation that significant variation in radio and optical emission occurs in periods less than one year, which implies that these radiations must be generated within a sphere of diameter less than 1 light year.

The situation with X-rays is even more astounding in that large regular fluctuations in the strengths of X-ray emission have been found with periods as short as 3 hours.

THE SOLAR MAXIMUM MISSION SATELLITE

The S.M.M. Satellite became operational in 1980, and was commissioned for solar observation. After a few months it became partially inoperative, but was repaired in 1984 using the space shuttle.

The satellite carried seven instruments including

a coronagraph/polarimeter	(C.I.P.)
an ultra violet spectrometer and polarimeter	(U.V.S.P.)
a soft X-ray polychromator	(X.R.P.)
a hard X-ray imaging spectrometer	(H.X.I.S.)
a hard X-ray burst spectrometer	(H.X.R.B.S.)
an X-ray spectrometer	(G.R.S.)
an active cavity radiometer irradiance monitor	(A.C.R.I.M.)

Among many findings are included the following. The C.I.P., as its name implies, is used to study the corona. It is capable of high-speed multicolour photometry and polarimetry (that is, measuring light intensity and degree of polarisation). Solar flare intensities using the C.I.P. have shown that solar flares must be driven largely by magnetic forces rather than by gas pressure gradients, as was previously assumed.

The U.V.S.P. has been used to measure the degree of polarisation due to magnetic fields, of an ultra violet spectral line of carbon IV, originating from material above, a large sunspot. The findings show that previous estimates of the magnetic field strengths associated with sunspot activity are much too low. Observations on the same spectral line also show significant oscillatory motions of matter just above sunspots.

177

The H.X.I.S. has been used to provide a complete evolutionary description, of the morphology and plasma characteristics, of a number of solar flares.

The G.R.S. is used to study the short-wavelength limit of solar radiation. At these energies the sun emits both a continuum and a number of X-ray lines. The latter are thought to arise from the interaction of fast ions with the denser portions of the solar outer atmosphere. Hence X-ray emission is an indicator of the presence of accelerated ions.

The A.C.R.I.M. is used to monitor the radiative flux of the sun. One particular investigation concerns the problem of what happens to the radiative energy 'missing' from sunspots, since it obviously does not emerge from the sunspots.

The short list of S.M.M. investigations given here represents only a small fraction of the total work which has been carried out using this satellite.

SATELLITE OBSERVATIONS IN THE NEAR FUTURE

A number of astronomical satellites have been approved and funded, and are scheduled to operate in the near future. These include the following.

The Space Telescope (1988/9 – N.A.S.A.)

The Space Telescope is a large Cassegrain optical telescope, with a 2.4 metre primary mirror. It is due to be placed in an orbit, 500 kilometres above the earth, using the space shuttle. It will have an operating life of 15 years, though options are available to operate it for many decades. All observations will be without the obscuring effects of the earth's atmosphere, which is the principal reason for building the telescope.

On board, image intensifiers will enable objects to be detected down to the 28th and 29th magnitude, which is about 5 magnitudes fainter than earth-bound telescopes can detect.

It will initially contain five detecting instruments, though more can be added at a later date using the space shuttle. These include a wide-field planetary camera, a faint-object camera, a faint-object spectrometer, a high-resolution spectrometer, and a high-speed photometer.

The wide-field planetary camera will provide high-resolution photographs of planets, and panoramic views of star fields. The faint-object spectrometer will be used to measure the chemical composition and dynamics of faint light sources, including comets, galaxies and quasars.

The high-resolution spectrometer will explore the physical characteristics of exploding galaxies and interstellar gas clouds.

The high-speed photometer will enable precise measurements to be made of the time variation of light intensity of pulsating and variable stars, which are important data for distance determinations.

The range of instrumentation available will allow observations to be made from 115 nm in the far ultra violet to 1 mm in the far infrared.

Hipparcos (1986 – E.S.A.)

Hipparcos will contain an optical system incorporating a 30 cm telescope, and will be used in the determination of star positions to an accuracy of 0.002 arc-seconds. Observations from earth are limited to an accuracy of about 0.1 arc-seconds.

O.S.S.-3 (1988/9 – N.A.S.A.)

This mission will operate a number of ultra violet telescopes. A spectrometer is included to measure wavelengths of less than 120 nm.

R.A.C.S.A.S. (1986 – U.S.S.R.)

This is the 'Radio Astronomy Cosmical System Aperture Synthesis'. It consists of a 30 m diameter dish of wire mesh, and is to be deployed in a low orbit of 500 km. It is to be used as part of a radio interferometer system, in conjunction with a 70 m dish in the Crimea.

R.O.S.A.T. (1987 – Germany/U.S./U.K.)

Built for X-ray observations, the satellite contains an 80 cm diameter, cylindrical, grazing incidence telescope. It is planned to survey the whole sky, perhaps cataloguing some 100 000 sources, and then concentrating on detailed investigation of some 10 000 sources.

ASTRO-C (1987 – Japan/U.K.)

A large detector built in the U.K. will further investigate X-ray spectra and the periodic fluctuations of some X-ray sources.

Long Baseline Interferometer (1987 – U.S.S.R.)

This project is a follow up to R.A.C.S.A.S. and comprises a 10 m dish which will be placed into an extremely high orbit of one million kilometres. It will again be used in conjunction with the dish in the Crimea, and will provide a resolution which is one-hundred times higher than is obtainable with ground-based interferometers.

Cooled Sub-millimetre Telescope (1987 – U.S.S.R./France)

This telescope comprises a one metre diameter mirror and will be used to study spectra at wavelengths of 0.1 mm to 2 mm. The telescope will be cooled to 40 K and the detector to 1 K.

Gamma-ray Observatory (1988 – N.A.S.A.)

This satellite will carry four X-ray detectors, three of which cover a wide energy range from 100 keV to 500 GeV, with high resolution, and the fourth will investigate spectral lines emitted by radioactive atoms ejected by supernovae.

Cosmic Background Explorer (1988 – N.A.S.A.)

This will provide a detailed study of the structure and spectrum of the microwave background radiation.

Infrared Space Observatory (1990 – E.S.A.)

I.S.O. is designed to take over where I.R.A.S. left off. In constructional details the two satellites are very similar, but whereas I.R.A.S. was a survey satellite and was used to catalogue a great many sources, I.S.O. is an 'observatory' and will provide an in-depth study of a smaller number of sources. To achieve this, I.S.O. must operate for a longer period than I.R.A.S. did. To prolong the rundown time of the coolant, I.S.O. will carry the 100 kg of superfluid helium at 3 K to cool the detectors and the telescope, as did I.R.A.S., but in addition will carry a further 50 kg of slightly warmer liquid hydrogen to provide the bulk of cooling for the satellite. The anticipated lifetime is between 1½ and 2½ years.

I.S.O. has three detector systems. A large semi-conductor (indium antimonide) 'chip' divided into 1024 radiation-sensitive segments can detect the source at wavelengths of 1 to 5 micrometres. It may be paired with a newly developed silicon–bismuth chip which will simultaneously record at 5 to 18 micrometres.

The second system is a photometer to measure the brightness of a source simultaneously at 12, 40 and 100 micrometres with three chips made of silicon-gallium, germanium-beryllium and germanium-gallium respectively.

The third system comprises a pair of spectrometers covering the range between 2 and 70 micrometres.

Bibliography

Astronomy, F. Hoyle (Macdonald)

Astronomy and Cosmology, F. Hoyle (W. H. Freeman)

Atoms, Stars and Nebulae, L. Goldberg and L. H. Aller (McGraw-Hill)

Early Solar Physics, A. J. Meadows (Pergamon)

Elementary Astronomy, O. Struve (Oxford University Press)

Focus on the Stars, H. Messel and S. T. Butler (H.E.B.)

'Frontiers in Astronomy', *Scientific American*

General and Comparative Physiology, W. S. Hoar (Prentice-Hall)

'Images and Information', *Open University O.U. ST. 291*

Introductory Astronomy and Astrophysics, Smith and Jacobs (Saunders)

'New Frontiers in Astronomy', *Scientific American*

Perspectives of Modern Physics, A. Beiser

Stellar Evolution, A. J. Meadows (Pergamon)

The Manual of Photography (Focus Press)

The Sun, G. Abetti (Faber and Faber)

The Sun and Stars, J. C. Brandt (McGraw-Hill)

The Sun and its Influence, M. A. Ellison (Routledge and Kegan Paul)

Full details of satellite observations are published in the scientific journals *Science* and *Nature*.

Index

R.R. Lyrae stars 113, 133
Rutherford–Appleton laboratory
    166

Sagittarius 149
satellites 165–81
Saturn 96
scattering of light 34, 63
secondary emission 80
secular parallax 113
silver bromide photographic process
    57
Sirius 19, 115
sky colour 36
'S.M.M.' (Solar Maximum Mission)
    satellite 177
solar flares 155, 177
solar noise 155
'Space Telescope' 178
spectra 9–14, 16, 32
spectral class 'A' stars 113
spectral classification 124, 126
spectral line broadening 106, 131
spectral sensitivity of
    photoemissive surface 75
    photographic film 68
    photomultiplier 82
    thallium sulphide cell 86
    the eye 51
spectrometer 5, 6, 168, 172
spectroradiometer 16, 17
spectroscopic binary systems 98,
    101, 102
spectroscopic parallax 144
spiral arms 133, 160, 175
statistical parallax 113
stellar clusters 113, 131, 145
stellar coronas 176
stellar evolution 140
stellar populations 'I' and 'II' 133–6,
    138, 147
Stephan's law 20, 21
sub-image 59
sun 37–43, 118, 138, 139, 153,
    155, 162, 170, 177
sunspots 40, 154, 177
supergiants 129, 130, 142, 148

supernova 42, 139, 143, 156, 158,
    167, 175
synchrotron radiation 153, 157,
    170

telescopes 149, 170
thermal Doppler broadening 106–9
thermal radiation 14, 15, 153,
    169
thermonuclear reactions 136,
    140
transmission limits of optical materials
    6
'transmission' of a negative 64
trigonometric parallax 110, 144
'triple-alpha' reaction 42, 139, 142
'twenty-one centimetre' radiation
    156, 158

'U.B.V.' system 119
'Uhuru' satellite 170, 172, 174
ultra violet catastrophe 23
ultra violet photography 72
ultra violet satellite observations
    167–9
universe, size and age 92

'velocity curves' for spectroscope
    binaries 101
visual magnitude 119
visual purple 47

water vapour 33
wave mechanics 30
Weber–Ffechner Law 55
'white dwarfs' 9, 12, 130, 142, 176
Wien's law 19
'work function' 73, 74
'W–Viginis' Cepheids 147

X-ray background radiation 174
X-ray photography 72
X-ray satellites 169–77
X-ray stars 172

ylem 41